Ninja
UNMASKING THE MYTH

STEPHEN TURNBULL

Frontline Books

NINJA
Unmasking the Myth

This edition published in 2017 by Frontline Books,
an imprint of Pen & Sword Books Ltd,
47 Church Street, Barnsley, S. Yorkshire, S70 2AS

ISBN: 978-1-47385-042-2

Pen & Sword Books Limited incorporates the imprints of Atlas,
Archaeology, Aviation, Discovery, Family History, Fiction, History, Maritime,
Military, Military Classics, Politics, Select, Transport, True Crime, Air World,
Frontline Publishing, Leo Cooper, Remember When, Seaforth Publishing,
The Praetorian Press, Wharncliffe Local History, Wharncliffe Transport,
Wharncliffe True Crime and White Owl.

For more information on our books, please visit
www.frontline-books.co
email info@frontline-books.com
or write to us at the above address.

Printed and bound by TJ International Ltd, Padstow, Cornwall
Typeset in 10.5/13 Palatino

Contents

Preface

I have had problems in the past with cats coming into my garden and making a mess, but last Christmas I was provided with a possible solution. It was the gift of a plastic garden gnome in the shape of a ninja. He now stands there as a menacing deterrent to feline infiltration, and it has been interesting to note that all visitors to my garden (including, I hope, the cats) identify the figure correctly as a ninja. When one considers that the word did not appear in any Japanese–English dictionary until 1974, such instant recognition is a measure of how familiar the concept has become in little over half a century. *Ninja: Unmasking the Myth* will tell the remarkable story of how this familiarity came about, and how a wide range of stories from Japan's past became transformed into a multi-million dollar cultural phenomenon that is based on the belief that there was once a time when people who dressed exactly like my garden gnome climbed into castles and set fire to them.

This belief lies at the heart of what I have termed 'the myth of the ninja'. It also poses a considerable dilemma for an author, because a choice has to be made between explaining the popular ninja figure and merely accepting him. A serious study should show how the ninja of the movies (whose skills at the martial arts sometimes encompass the superhuman ability to fly) emerged from the realities of the past, but most popular books do not treat the topic in this way. Instead they carelessly accept fantasy as reality and retell authentic historical accounts of Japanese undercover warfare as if they were actually performed by these comic book characters, and I must confess that I have written both types of book.

First of all came *Ninja: The True Story of Japan's Secret Warrior Cult* in 1991, a serious if flawed attempt to discover the reality behind the ninja by using original Japanese sources.[1] I then wrote *Ninja AD 1460–1650* for the Osprey *Warrior* series in 2003, in which specially commissioned colour plates showed genuine episodes from history being performed by my 'garden gnome' ninja dressed in black, whom I sent into battle again in 2008 for the children's book *Real Ninja*. To make matters worse, my next project, *Ninja: The (Unofficial) Secret Manual* will be a tongue-in-cheek handbook for ninja training written in the style of a seventeenth century master of ninjutsu.[2] My loyal readers would certainly be forgiven if they were to shout, 'Is the ninja real or not? Make your mind up!'

I am, however, not the only one to have a foot planted firmly in each camp, because Japan's leading authority on the ninja phenomenon has recently done the same. Mie Prefecture in central Japan includes the former province of Iga where the ninja idea first took root, and in 2012 Professor Yamada Yūji of Mie University's Faculty of Humanities, Law and Economics risked academic scorn by setting up a research project called the Iga Ninja Culture Collaborative Field Project to investigate the lively cultural phenomenon on his own doorstep. The effort was followed in 2017 by the launch of Mie University's International Ninja Research Centre, for which I had the honour of delivering the inaugural lecture.[3] The initiative is already bearing fruit, with the mounting of conferences, public lectures and the creation of a database of ninja-related resources.[4] It has also led to an increasing number of publications by people associated with the project, including Yamada's own *Ninja no Rekishi* (A History of Ninja).[5] *Ninja no Rekishi* is a work of scholarship that draws on much of the same source material that is translated here, but Yamada has also produced his own 'ninja training manual' in which every aspect of the ninja is played straight for a young audience with lively cartoon illustrations that include the notorious ninja water spiders conveying their wearer across a stream. His most recent work, *Ninja ninjutsu chō hiden zukan*, is even more outrageously commercial because it portrays an indiscriminate mixture of historical and fictional characters as manga ninja figures.[6]

Nevertheless, both Yamada and I encourage the reader to explore more widely and seek out the truth behind the fantasy, unlike some authors who refuse to accept historical realities at all and produce supposedly serious books for adults that have more in common with the fantasy training manuals. In these books ninja are shoehorned into every conceivable historical operation that involves the slightest element of secrecy, and it is done in a way which suggests that the sources for the claims are revealed truths unearthed from secret scrolls.

One of the greatest contributions made by the Mie University initiative has been the identification of much previously unknown source material. This has revolutionised my own understanding of the subject, which now bears no relation to the situation in 1991. For my *Ninja: The True Story of Japan's Secret Warrior Cult* I translated the historical accounts that were before me and interpreted them as I understood them at the time. My translations were largely accurate, but in the absence of the modern research now available I unfortunately allowed myself to become over-dependent on works such as Yamaguchi's imaginative *Ninja no Seikatsu*, in which almost any secret operation in Japanese history is credited to a ninja.[7] I had long intended to produce a revised edition of *Ninja: The True Story of Japan's Secret Warrior Cult* because I was quite sure that one day someone would mount a serious challenge in writing. That time has now come and I am pleased that I am the one putting that challenge into words, although so much has needed to be revised that only this completely new book will suffice. *Ninja: Unmasking the Myth* will also extend the earlier book's remit by looking at the historical and cultural phenomenon of the ninja as a whole, thereby providing some clues as to how it is possible to write contradictory books about ninja with an easy conscience.

I could not have produced this book without the personal cooperation of the staff of Mie University, and I would first like to thank Guillaume Lemagnen, who introduced me to their research team in 2014.[8] Professor Yamada then made me very welcome and placed the department's resources at my disposal, not the least of which was supplying me with a copy of the very rare 1956 booklet *Ninjutsu* by Okuse Heishichirō, the first post-war publication on the topic of ninja. Such information helped greatly in planning very fruitful field trips between 2014 and 2017, some of which were carried out with the help of members of staff.

Two other people have been particularly helpful in the field of the ninja as a cultural phenomenon. Jonathan Clements' academic perspective on movies, martial arts, manga (graphic novels) and anime (animated films) opened many doors for me about the ninja as a fictional character.[9] I have also gained a great deal from Keith Rainville's brilliant and highly detailed website 'Vintage Ninja', which approaches the ninja from the point of view of popular culture. His collection of movie references and ninja ephemera from trading cards to political propaganda is unique and extensive, and I recommend this source wholeheartedly.[10] Some of the ideas in *Ninja: Unmasking the Myth* were first aired in an article based on a lecture I gave at Kennesaw State University in Atlanta, Georgia in 2013.[11] The paper has been widely circulated and I am genuinely grateful for the comments made by both scholars and ninja enthusiasts. In particular I acknowledge the contributions from Kawakami Jinichi and Ikeda Hiroshi. Antony Cummins' generous comments have also been very helpful. His translations of the seventeenth century ninja manuals also provided me with a useful starting point for their study. [12]

Above all I wish to thank my dear wife Marlene, whose unwavering support sustains me throughout all my writing. In 2017 she accompanied me on a research trip to Japan and was able to experience at first hand the phenomenon of the ninja that has fascinated me for so long. I warmly welcome input from anyone who likes to get involved with ninja (apart from cats).

Stephen Turnbull

Author's Note
As part of the argument over the ninja's authenticity relates to the use of language I have introduced Japanese characters into the text where it is necessary for the clarification of nomenclature or certain technical terms. They are also added to the titles of certain key documents, personal names and places where I feel they are necessary. I have also retained the Japanese convention of presenting Japanese personal names with the surname first.

Chapter 1

The Universal Ninja

The ninja has become a familiar figure in Japanese popular culture as the world's greatest exponent of undercover warfare. As a masked secret agent dressed all in black he infiltrates castles, gathers vital intelligence and wields a deadly knife in the dark. He possesses almost magical martial powers and is capable of extraordinary feats of daring. The ninja is associated with two specific areas of Japan called Iga and Kōka, from where he sells his services as a mercenary, and when in action his unique abilities include confusing enemies by making mystical hand gestures or by sending sharp iron stars spinning towards them.

This is the exciting image that has been enjoyed in books, films and television series for over half a century. In recent years it has also become an official part of 'Cool Japan', an initiative launched in 2010 by Japan's Ministry of Economy, Trade and Industry to promote Japan's creative industries to foreign countries through the use of popular culture, and a great number of people believe, or want to believe, that the 'cool ninja' is based on solid historical reality.[1] From their point of view the man behind the mask is no more than a lively modern manifestation of an unbroken tradition dating back to a time when specialist ninja warriors really did climb into castles using specialist techniques called ninjutsu.

The topic of ninjutsu, which has a mythology all of its own, will also be covered in this book. It is conventionally understood to mean what the ninja did, although an examination of its historical usage reveals that it once meant sorcery or magic, not techniques subject to human limitations. Those esoteric definitions have long been abandoned by certain individuals nowadays who claim to practise ninjutsu as a

1

martial art in its own right. Just like the ninja themselves, the martial art of ninjutsu is supposed to have its own authentic history, which was miraculously preserved, recorded arcanely in secret scrolls and then passed on to carefully selected modern practitioners who staunchly defend its historical authenticity in a way that is reminiscent of the passions displayed by the members of a religious cult. Yet even the most devoted fans will acknowledge that a certain amount of exaggeration must have taken place to produce the fantasy figure of today and his related behaviour. Human beings, after all, cannot escape from combat by flying backwards on to a roof, so the argument becomes instead one of identifying a genuine continuity with what may have existed in the past. On the opposite side of the debate stand the dogged ninja skeptics, yet even they tend to stop short of declaring that the idea is a total fabrication. The usual approach is simply to accept the ninja and his ninjutsu as genuine historical phenomena that have long been greatly romanticised and highly commercialised.

As this book will show, the romanticisation of the Japanese undercover warrior goes back many centuries, so for readers unfamiliar with Japanese history a short outline may be needed at this stage. Pre-modern Japan was characterised by sporadic and confusing civil wars in which espionage and undercover warfare inevitably played a role. The conflicts reached a peak between the mid-fifteenth and the early seventeenth century, an era customarily compared to Ancient China's Warring States Period and therefore given that name: *Sengoku Jidai*, the Sengoku Period or 'Age of the Country at War'. To understand why there was such disorder at that time we must look back a few hundred years to the Gempei War of 1180–85, a conflict that had pitted two samurai families against each other in a major war for the first time in Japanese history. The Gempei War led to the reduction of imperial power and the establishment of a permanent form of the once-temporary position of Shogun, who proceeded to rule Japan as a military dictator through and on behalf of the samurai class. The system lasted until 1868, although a severe challenge was made to the Shogun's authority by the Ōnin War of 1467–77. That tragic conflict began when a succession dispute within the Shogun's own family led to fighting in Kyoto, then Japan's capital. Much of Kyoto was devastated, and when the conflict spread to the provinces the Shogun was powerless to stop it. The Ōnin War was the start of the Sengoku Period, and it is during that time that we read of rival *daimyō* 大名 (local lords, literally 'big names') using undercover operations against each other.

2

The eventual reunification of Japan began with the military genius called Oda Nobunaga (1534–1582) during the 1560s. It was completed under his equally talented successor Toyotomi Hideyoshi (1536–1598) and consolidated under Tokugawa Ieyasu (1542–1616). The reunification process was largely a military operation that was enforced by land surveys, the transfer of landowners and forcible disarmament. The ultimate result was a shift from the rule by provincial *daimyō* to a national hegemony where the only *daimyō* left were the appointees of the national ruler. Following Ieyasu's triumph at the battle of Sekigahara in 1600 the Tokugawa family re-established the Shogunate in 1603 and continued to rule Japan until the Meiji Restoration of 1868. Their era was called the Tokugawa or Edo Period, from the Shogun's capital of Edo, the city we now know as Tokyo. The civil warfare of the Sengoku Period had been replaced by a long time of peace sustained by severe martial law and supported by an extensive state intelligence network that would also contribute to the concept of a ninja.

The notion of an unbroken continuity between a number of well-recorded secret operations associated with these historical events and the ninja and ninjutsu of today constitutes the essence of the ninja myth, and it goes far beyond the retelling of ancient stories. It is a very complex entity that contains many elements of the classic notion of an invented tradition, a concept from the social sciences that will be discussed in detail later, although the ninja myth appears to differ from the usual model of an invented tradition because of its long history and its dynamic nature.

The building blocks for the ninja myth can easily be traced back to the civil wars of the Sengoku Period, although some enthusiasts will go back even further in time and find links in the mists of imperial antiquity, seizing upon certain historical figures and projecting the idea backwards to credit them with being ninja. The legendary Prince Yamato Takeru resorts to subterfuge on at least one occasion, including dressing up as a woman, making him the forerunner of the ninja in some eyes.[2] Kusunoki Masashige (1294–1336), lauded for his devotion to the emperor and the greatest hero in *Taihei ki*, the chronicle of the Nanbokuchō civil wars of the fourteenth century, supposedly used guerrilla tactics including booby traps and dummy warriors, but one should not claim that these activities prove that Masashige was the founder of a specific *ryū* 流 (school or tradition) of ninjutsu without much more supporting evidence.[3] Even more contrived is the story of the rebellious sorcerer Fujiwara Chikata, who conjured up four devils to overcome an emperor. Chikata's devils

3

have been identified as ninja by some writers, but this is just a further example of the 'retrofit' notion of a ninja.[4]

It is also important to note at this stage that in addition to misconceptions about who the ninja's historical antecedents were, there are several misunderstandings about what they did. As the following pages will show, the historical forebears of the ninja were infiltrators and spies, not assassins, and only one pre-modern document lists assassination among their possible roles.[5] Nevertheless, many stories of assassination have fed into the ninja myth. For example, there is the exciting account in *Taihei ki* of the murder of Honma Saburō by the youth Kumawaka, who escaped by climbing up a bamboo trunk (in itself no mean feat) and allowing it to deposit him in a place of safety. It is a good story that may with complete justification be described as a 'ninja-like assassination', but the boy was a revengeful opportunist, not a trained infiltrator.[6]

Misinterpretations like these demonstrate that instead of a smooth continuity between past and present, the ninja myth has suffered breaks in both flow and concept, which have been bridged in retrospect by huge leaps worthy of the most acrobatic ninja from the movies. The evidence for this will be provided from a wide range of historical and literary sources, but the first step will be to tease out the essential features of the archetypal ninja as he is presently understood to provide a baseline for comparison with what may have existed in the past. The second stage of enquiry will be to look at accounts of undercover warfare that were recorded by unbiased eyewitnesses when wars were still taking place. The third stage will be to examine the wide range of written material produced after Japan's wars had ceased. We will see how during the peaceful Tokugawa Period the earlier descriptions were exaggerated and manipulated within a wide range of literary works to produce what would become the building blocks for the ninja and ninjutsu of today. The book will continue with a detailed examination of how the ninja emerged within pulp fiction and films in post-war Japan and went on to become a worldwide cultural phenomenon. It will conclude with the ninja's extraordinary new role during the run-up to the 2020 Tokyo Olympics as a physical and moral exemplar for the nation's youth.

The benchmark ninja in word and deed
An extremely complex reality lies behind the origins and development of the ninja myth, although its complicated nature is obscured by the

remarkably consistent image of a ninja that emerged during the 1950s and 1960s and has persisted to our own day. As noted earlier, this archetypal ninja figure is a secret agent dressed from head to foot in a tight-fitting black costume who deploys a unique armoury of weapons in his superlative practice of ninjutsu. These esoteric techniques include using an unusually straight sword as a climbing device, throwing star-shaped *shūriken* (the iron 'ninja stars' spun from the hand), scattering caltrops and discharging smoke bombs to cover his withdrawal. In this he is sustained by mystical powers acquired by making esoteric hand gestures. The ninja is a key player in Japan's civil wars who crosses moats, scales walls and climbs into castles to acquire intelligence, to cause havoc among the garrison or to carry out an arson attack. His services are hired from the ninja masters of Iga and Kōka, a tiny area of Japan from where all ninja excellence ultimately derives.

So far, so familiar, but popular culture has also placed him in a particular social environment of impoverished lower-class part-time samurai-farmers, in which he can be either the 'good guy' or the 'bad guy'. The good ninja of the film world is a militarily élite yet socially inferior mercenary who is a member of a hierarchical ninja clan and augments this Confucian notion by a deeply spiritual side that has links to esoteric Buddhism. His clan consists of rugged proletarian warriors who fight oppression, although their rivals may sometimes be another ninja clan who have been seduced into serving a wicked overlord. Those ninja are always the bad guys, and the worst of them all belong to mysterious clandestine organisations which the heroes must overcome. Sometimes the places called Iga and Kōka are transformed from being mere geographical locations into rival clans, schools or even secret societies.

The world of both good and bad ninja can be dark and violent, although the action tends to stay within physical limitations, give or take a few somersaults. The benchmark black-clad ninja is therefore essentially human, so it is not at all difficult to identify parallels between what ninja do in the movies and what certainly went on many centuries ago. Yet fantasy is never far away, because there appears to have long been a tacit understanding that if the benchmark ninja is a creature of the imagination, then the addition of a few superhuman elements will be fully acceptable to one's readers and viewers. Thus it was that in 1958 Yamada Fūtarō's influential novel *Kōka Ninpōchō* created a painstakingly accurate historical environment in which superhuman ninja could transform their hair into porcupine quills and use them as weapons.[7]

5

A different development has been to abandon the visual shorthand of the benchmark ninja's black costume for a punk outfit or a space suit to produce a completely outrageous figure who suffers no restrictions, physical or otherwise. One example is the character called Naruto from the manga series of the same name who dresses in a bright orange jumpsuit. He is referred to as a ninja, but all the old visual clues have disappeared completely. This boundary-free concept may also be encountered in a safer junior version: the *kawaii* ('cute') ninja. The children's ninja hero is a brave but highly sanitised fantasy warrior who is always the good guy. These cute little ninja are very accomplished at fighting opponents who are as likely to be monsters and wizards as brutal overlords. They use magic as much as martial skills and will readily swap their black costumes for bright colours. Such a character is exemplified by the bespectacled boy hero Nintama Kentarō, who may be encountered alongside the archetypal man in black in studio theme parks and ninja villages, where all the family can dress up in pink or blue ninja outfits and share in the fun.[8]

The other consistent element of the ninja as he is currently understood lies in the choice of language for his name, which is written using the two *kanji* (Chinese-derived ideographs) 忍 and 者. There is no problem over the meaning of the second character. *Sha* (or *ja*) 者 simply refers to a practitioner of something, as in the word *geisha* 芸者. The word ninjutsu 忍術 uses a second character 術 meaning techniques, but the first character *nin* 忍 in both words can have two very different meanings. It is usually taken to mean secrecy, invisibility or concealment, but its primary definition involves the idea of endurance. For example, *The New Nelson Japanese–English Character Dictionary* gives the primary meaning as 'bear, endure, put up with' and the meaning as 'hiding' only appears in second place.[9] In the original Chinese *ren*, the Mandarin pronunciation of *nin* has no association with stealth at all and instead appears in dictionaries as 'endure; tolerate; put up with'. Examples of its use in modern Mandarin include 'Don't lose your temper over trivial matters', and none of the coinages presented in Manser's Chinese Dictionary have anything to do with stealth.[10] *Ren* also appears in this form in Axel Schuessler's *ABC Etymological Dictionary of Old Chinese*, in which it is explained that it was pronounced *nin* under the later Han Dynasty. It is listed in Schuessler's dictionary with two meanings of 'to endure' and 'to be cruel', but there is again no mention of secrets or secrecy.[11] In fact *ren* seemingly does not refer to stealth in any Chinese context until the translation of

Japanese books back into Chinese. In Japanese *taeru* (堪える or 耐える) is another verb that means to endure. When its root is combined with *nin* it becomes the noun *kannin* 堪忍 (endurance) and has a different verbal form as *tae shinobu* 耐え忍ぶ (to endure). The two characters also appear in the noun *nintai* 忍耐, which means patience, perseverance and fortitude.

Later chapters will examine the very important conclusions about ninja and ninjutsu that have arisen when the character *nin* is regarded as signifying endurance instead of stealth. For now we will note that the *nin* in the ninja of the movies always indicates secrecy. He is a man of mystery rather than someone known for his patience or fortitude. There is, however, a further complication to consider, because most ideographs in the Japanese language can be read in two different ways. The first is the *on* reading, which is the modern approximation of the original Chinese pronunciation when the ideograph was introduced from Chinese. The *kun* reading is the pronunciation of the native Japanese word that had the same meaning as the introduced character. Sometimes, but not always, an ambiguous or unfamiliar set of ideographs will have small *furigana* (phonetic syllables) printed next to them to indicate the correct reading. In the case of 忍者 *nin* and *sha* are *on* readings, while the *kun* readings are shinobi and mono respectively. The two characters are sometimes separated by the phonetic particle *no* の, otherwise 'no' is just implied to identify the 'practitioner of secrecy' in its *kun* reading as a shinobi no mono or just a shinobi. Similarly, 忍術 can also be read as shinobi no jutsu, although ninjutsu appears to have been the preferred reading throughout most of its written history.

Taking one's clues solely from the addition of *furigana* to Japanese texts, a fully confirmed *on* reading of 忍者 as ninja does not appear in any written source until 1955 when it emerges in the novel *Sarutobi Sasuke* by Shibata Renzaburō, where 忍者 is glossed to be read in that precise way on its first appearance.[12] Shinobi (no mono) should therefore be taken as the general rule until the late 1950s, although there are some exceptions. For example, in his *Gendaijin no Ninjutsu* of 1937 Itō Gingetsu, an important early writer on the subject of ninjutsu, used the word 忍者 but glossed it as *ninsha*, not shinobi no mono.[13] There are also exceptions after 1955. In 1957 a seventeen-volume manga series was published that used ninja in its title as far as bibliographic references were concerned, but the title *Ninja bugeichō* (忍者武芸帳) is not glossed, and to complicate matters further there is a mention of 忍者 in one frame with the word glossed as shinobi no mono.[14]

What may be more surprising to the general reader is 忍者's complete absence outside Japan until the 1960s under any reading: ninja, ninsha or shinobi no mono, as can be demonstrated by tracing its first appearance in Japanese–English dictionaries. The compound is missing from the dictionaries by Takahashi (1879), Brinkley (1896), Hepburn (1907), Inouye (1920), Saitō (1927), Rose-Innes (1942), Vaccari (1949) and Kenkyūsha (1954). It is included for the first time in Nelson's dictionary of 1962 using Itō's reading of it as 'ninsha (ancient) spy' together with 'shino(bi) mono spy, scout', and retains that form until the second revised edition of 1974 when ninsha finally becomes ninja. Also in 1974 the revised Kenkyūsha introduced the word ninja for the first time as 'a samurai who mastered the art of making oneself invisible through some artifice and chiefly engaged in espionage activities'.

In marked contrast to the name of its practitioners, the word for the ninja's techniques appears to be a well-known expression to foreigners right from the end of the nineteenth century, although ninjutsu had a very different meaning from the one it enjoys today. There were no hints then that it might ever indicate a martial art. Instead it appears in Brinkley (1896) as 'the magical art of concealing one's self from another's sight'. In Inouye (1920), Saitō (1927), Vaccari (1949) and Kenkyūsha (1954) ninjutsu is similarly defined as 'the art of invisibility'. Nelson (1962) has it as 'occult art', adding 'the art of making oneself invisible' to the definition in 1997. There is also no suggestion at that stage that it might refer to spying techniques. That idea took time to be disseminated abroad, even though it was already well established in Japan. In 1955 the authoritative dictionary Kōjien had defined ninjutsu as 'one variety of mittei jutsu 密偵術 (the techniques of spying)' and linked it to espionage during the time of the samurai, adding that the Iga and Kōka areas were famous for it.[15]

To some extent the reading of 忍者 as either ninja or shinobi no mono is of more interest to Westerners than to native Japanese speakers, because to the latter the pronunciation of the word is less important than what it signifies. To Japanese enthusiasts ninja and shinobi are easily interchangeable. A ninja film released in 2017 bore the title *Shinobi no Kuni* ('Province of the Shinobi') rather than 'Ninja no Kuni', while one of Japan's newest ninja tourist traps is called the Shinobi no Sato (Shinobi Village) and only uses the word ninja in its English language publicity materials.

Whatever use may be made of it in Japan, I believe that the reading as ninja has greatly helped the overseas transmission of the idea simply because it trips more readily off the Western tongue. The shinobi therefore crossed an important cultural divide when they became ninja, although in recent years there has been an interesting tendency among Western ninja enthusiasts to prefer the term shinobi for their heroes on the grounds that its use in historical accounts confirms the authenticity of their belief in a long and genuine tradition. A contrast is therefore made between shinobi and ninja, whereby the vital continuity with the past that is essential to the ninja myth is vested in the supposedly historical shinobi. The word ninja is then reserved for the exaggerated popular development found in comic books and movies. In other words, the fantasy ninja can fly while the historical shinobi has his feet planted firmly on the ground and everyone is happy. Kawakami Jinichi, the Honorary Director of the Iga-ryū Ninja Museum, once summed up this point for me very neatly by defining a ninja as a fictionalised shinobi.[16]

No matter how acceptable this simple distinction between fictional ninja and historical shinobi may be to enthusiasts, it has never caught on with the wider public and is in any case readily ignored if there is a need to use the word ninja to make money. The official handbook that accompanied a major ninja exhibition for children in Tokyo in 2016 freely used the term ninja rather than shinobi when describing historical events.[17] Authors can always add a suitable explanation if their consciences are troubling them, and part of that explanation involves noting that the fact/fiction distinction has no real historical justification, because even if the benchmark ninja in the movies was the product of someone's imagination, there is also much that is fictional about the supposedly historical shinobi.

The first point to note in this context is that shinobi is not always the expression of choice where one might expect to find it. In works written before the 1950s we often find references to various 'ninjutsu practitioners' instead of 'shinobi', with the operative in question being called a *ninjutsu-sha* 忍術者 ('ninjutsu man') or a *ninjutsu-tsukai* 忍術使 い ('ninjutsu user'). In 1914, a time when ninjutsu still tended to mean sorcery, a character called Sarutobi Sasuke made his first appearance in a novel as a *ninjutsu no daimeijin* 忍術の大名人 (a great name in ninjutsu).[18] His adventures involved martial arts, deception and trickery, so modern commentators often carelessly refer to the book as the first ninja novel, even though an examination of the original shows that the word ninja was never used and Sarutobi Sasuke did not acquire the title of ninja until a further novel about him was written in 1955.[19] In 1937 the influential Itō Gingetsu preferred the expression *ninjutsu-sha* over shinobi no mono.[20] In 1957 Adachi Ken'ichi used *ninjutsu mono* in the first full-length book on the subject in the post-war period, and as late as 1963 Okuse Heishichirō, a man whose ideas about ninja and ninjutsu were fundamental in the development of the concepts, was still very sparing in his use of the words ninja or shinobi and preferred to use *ninjutsu-tsukai*.[21]

A further complication is revealed when one examines reliable historical accounts of undercover warfare dating from the sixteenth century or earlier, because in these writings the word shinobi can fulfil the functions of both an adverb and a noun and the two meanings are not usually related. The adverbial form (discussed further below) is essentially a general expression meaning 'in secret'. The noun form of shinobi is far more specific and was definitely one of several words used in the past to identify a person who carried out secret military operations such as spying. An important and non-partisan piece of evidence for this comes from an unexpected source: *Vocabulario da Lingoa de Iapam*, the Japanese–Portuguese dictionary produced by the Jesuit mission in Nagasaki in 1603. Shinobi is included there under the form of *xinobi*. It is written in this romanised way without any ideographs and the word is defined as a spy who sneaked into castles to gain intelligence. There are two supplementary entries: *xinobiuo suru* (to investigate) and *xinobiga itta* (to enter as a spy). The inclusion of shinobi in this important historical work confirms beyond all doubt that the Japanese word for spy, regardless of how it was written, was indeed pronounced 'shinobi' in 1603.[22]

Any further conclusions that may be reached concerning the origins or degree of specialisation of these shinobi spies will be discussed later, but it is important to note that the meaning of the word was to change rapidly with the establishment of the Tokugawa Shogunate in that same year of 1603. Numerous references to certain people called shinobi then began to appear, but the individuals so named were not necessarily doing the sort of things that we now associate with ninja. The official government shinobi of the Tokugawa Period investigated possible dissent under the unpleasant reality of martial law by means of intelligence work that did not involve the romantic business of climbing into a castle by night. They were engaged instead in sordid tasks such as listening to gossip through keyholes, ready to denounce their victims and give them over to torture and confession.

Nevertheless, as its inclusion in the Jesuits' *Vocabulario* confirms, by 1603 the word shinobi definitely meant a spy, and its first appearance in this vein would appear to have been in the fourteenth century *Taihei ki*, which tells us that on the fifth day of the seventh lunar month of 1338, a night of wind and rain, raiders described as *ichimotsu no shinobi* 逸物 の忍び burned down a defended shrine on Yahatayama.[23] The phrase is somewhat ambiguous because it could mean 'first rate shinobi' as a noun or just '[people] ...with shinobi skills'. Commenting on the passage, Yamada identifies the shrine as the Iwashimizu Hachiman Shrine to the southwest of Kyoto, and notes that in accounts of the time it is unusual to read about raiders carrying out such operations on rainy or windy nights for fear of losing contact with each other. The success of this attack, he believes, led to the idea that efficient operations of this type were performed by specialists of some sort, for whom the word shinobi would be used.[24]

The above accounts confirm the use of shinobi as a noun meaning a spy or an infiltrator. Further examples will be presented later, but the overwhelming use of the word shinobi in contemporary descriptions of undercover warfare is as an adverb, not a noun. It describes how something is done, not who does it. For example Yamada notes that in *Moromori ki* 師守記, a diary written at the time of the Nanbokuchō Wars, the entry for Enbun 5 (1360) 5th month 8th day describes how 300 men under Wada Masauji mount a night attack on the Yūki family's castle after approaching in secret. Wada's men are not called shinobi, just *tsuwamono* 兵 (soldiers or warriors) and the character *nin* appears only within the phrase that describes their secret arrival.[25]

That is the general rule. When 忍び appears in contemporary accounts of secret operations there is usually no overt intent to identify either a person or a class of people by using a noun. It is instead an adverb and is encountered most frequently in the context of an attack on a castle or a defended place as *shinobi iru* 忍び入る or *shinobi komu* 忍び 込む, both of which mean to enter in secret. Two excellent examples occur in *Shinchō-Kō ki*, the biography of Oda Nobunaga that was written by one of his own followers. *Shinchō-Kō ki* records how in 1579, 'in the middle of the night, Hashiba Chikuzen-no-Kami Hideyoshi stole into (*shinobi itte* 忍び入って) the fort of Kaizōji in Harima Province and seized it. As a result, the next day the enemy abandoned the nearby castle of Ōgō as well'.[26] Hashiba Chikuzen-no-Kami is of course the future Toyotomi Hideyoshi, who would one day reunify Japan after a series of huge military campaigns. He was known for his clever tactics and use of surprise, but here the future ruler of Japan is merely acting 'in a shinobi way'; he is certainly not being identified as a shinobi. The interpretation of shinobi in a purely adverbial sense is reinforced later in the same work in a somewhat amusing manner because it is used to describe a clandestine movement in the opposite direction when Araki Muneshige is forced by Nobunaga to flee from his castle in secret. 'On the night of the 2nd day of the 9th Month, Araki Settsu no kami slipped out 忍び出て(shinobi dete) of Itami'. Once again the protagonist is no shinobi. Araki is running away![27]

As noted above, the secret operatives in similar accounts are sometimes called shinobi, but if the word is used as an adverb the protagonists are more likely to be identified using simple words for a soldier such as the above *tsuwamono* or more specific terms such as *kanchō* 間諜 (spies) or *teisatsu* 偵察 (scouts). A few colourful metaphors also appear. *Kusa* 草 (grass), used by the Hōjō family as an alternative to shinobi, indicates that their own intelligence gatherers hid in the long grass.[28] *Nokizaru* 軒猿 (eaves monkeys) compares a spy to a monkey hiding in the eaves of a house.[29] According to *Kōyō Gunkan*, immediately prior to the fourth battle of Kawanakajima in 1561 Takeda Shingen sent *nokizaru* under the command of his celebrated strategist Yamamoto Kansuke to spy on the Uesugi. They reported back that Shingen should be vigilant and place the river between him and his enemies.[30]

Many other names occur too, among which ideographs that otherwise mean 'thief' or 'robber' are frequently found. Their important significance will be discussed later, but when modern authors in search of ninja recount these stories the original characters are often glossed to be read phonetically as shinobi or the ideographs themselves may even be replaced by 忍者. A good example is provided by the treatment given in modern times to *Arayama kassen ki* (the chronicle of the battle of Arayama). *Arayama kassen ki* is undated, but the fact that it was incorporated almost in its entirety into *Taikō ki*, the biography of Toyotomi Hideyoshi by Oze Hoan (1564–1640) that was published in 1626, means that it must date from before about 1620.[31] In the original version a raid is conducted on behalf of Maeda Toshiie in 1582 by *Iga no nusumi-gumi* 伊賀ノ偸組, literally a 'band of robbers from Iga'.[32] In Yoshida's rendering of *Taikō ki* into modern Japanese the expression 'Iga no nusumi-gumi' has been replaced by 'Iga no shinobi-gumi'.[33] This is also how it appears in Iguchi Asao's biography of Maeda Toshiie.[34] Both are typical examples of how the ninja myth has taken hold to turn a gang of thieves into heroic specialist super-samurai called shinobi. Needless to say, popular books about the same events usually give the concept a further twist and replace the word shinobi with ninja.

The ninja and the invention of tradition

These alternative interpretations of the word shinobi illustrate better than almost anything else that the ninja myth depends on an act of faith in some form of continuity between the fantasy masked man of today and the historical shinobi of the past, but one huge obstacle lies in the way of the serious researcher who wishes to prove the theory either way. That obstacle is Japan's Tokugawa Period, because the authors of the Age of Peace were particularly adept at exaggerating what had gone before, and that included shinobi. As memories faded during the mid-seventeenth century and fighting was replaced by peacetime soldiering and bureaucratic duties, various authors tried to keep alive the traditional martial virtues. In spite of a common understanding of contemporary shinobi as official government spies, the writers, and to a lesser extent the artists, of the time were more than willing to lend a retrospective glamour to their name in newly written war chronicles and military manuals.

When undercover warfare (in its widest sense) is encountered in these works we find genuine exploits, equipment and their underlying philosophy being retold and often embellished for the edification of readers who were too young to have known anything of war but now ruled Japan as members of a military aristocracy. The two essential movements towards the ninja myth that are found in these accounts involve an increased use of the word shinobi in its noun form and the suggestion that the areas of Iga and Kōka specialised in undercover activities. These basic 'ninja elements' are noticeable from very early on in the martial literature of the Tokugawa Period, of which the most important example is *Iran ki*, the tub-thumping chronicle of the Iga Rebellion of 1579–80 written in 1679 by a local patriot called Kikuoka Jōgen. In a similar vein, *Ōmi Onkoroku* of 1684–88 describes the battle of Magari of 1487 when the Shogun Ashikaga Yoshihisa suffered guerrilla attacks by men from Kōka as the first operation by 'the Iga-Kōka *shinobi no shū* who are spoken of highly throughout the world'.[35] Several of the serious Tokugawa military manuals also have sections about undercover warfare in which can be found many hints of what was to come later. *Buyo Benryaku* 武用弁略 of 1684 contains a section on espionage that begins with the words, '忍者 (the expression is glossed as shinobi no mono) hid themselves in their own and other provinces and sneaked into enemy castles, strong though they were…. From ancient times the men of Iga and Kōka were distinguished in this and passed their knowledge on from generation to generation.'[36] In 1806 another military

13

manual called *Buke Myōmokushō* 武家名目抄 explains deep within its 381 volumes that shinobi were spies who carried out broadly similar tasks.[37]

An overall pattern is clearly developing, and by the end of the Tokugawa Period descriptions of undercover operations that happened centuries earlier moving rapidly towards the concept of a ninja as we now understand it. By 1833 one brave samurai simply described in 1698 as someone 'with shinobi skills' has become a 'shinobi no mono'.[38] Finally, Narushima Ryōjō's *Nochi Kagami* 後鏡, a history of the Ashikaga shoguns published in 1853, presents the idea that Japan's spies were linked to Ancient China through some imagined international brotherhood along with a reference to the battle of Magari, while its careless mention of the fictional character Tobi Katō makes the work sound even more like a popular ninja book written during the 1970s:

> they were said to be men from Iga and Kōka. In order to acquire secret intelligence they sneaked freely into enemy castles where they were assumed to be friendly troops. In the lands of the West (China) they were called *saisaku* 細作. Strategists called them *kagimono hiki* かぎ物ひき ('sniffers and listeners'). Round about time of the Eiroku Era there was one called Tobi Katō, a great expert. As for their name becoming famous throughout the whole of the country, this was since the time of their service against the Shogun Yoshihisa when he attacked Rokkaku Takayori during the Magari campaign. The family of Kawai Aki no kami served at Magari and were pre-eminent in front of the Shogunate's great army. Since then the 伊賀者 Iga-mono (men of Iga) have been admired for generations and this name Iga-mono was established.[39]

At about the same time pictures began to appear showing men dressed in black carrying out secret missions such as assassinations. Illustrations and written accounts such as these would provide further building blocks for the ninja myth, and their existence also shows that new elements have been sporadically introduced to the concept over a period of about four centuries. This is an exciting observation, because it means that the ninja myth may provide a fascinating and probably unique example of Hobsbawm and Ranger's classic notion of 'the invention of tradition', which they defined as:

> a set of practices, normally governed by overtly or tacitly accepted rules and of a ritual or symbolic nature, which seeks to inculcate

certain values and norms of behaviour by repetition, which automatically implies continuity with the past. In fact, where possible, they normally attempt to establish continuity with a suitable historic past.[40]

The invention of tradition is not necessarily a pejorative term. None of the examples given in Hobsbawm and Ranger's book have been dreamt up from nothing, and Hobsbawm notes that, 'There is probably no time and place with which historians are concerned that has not seen the "invention" of tradition in this sense'. He also identifies 'the use of ancient materials to construct invented traditions of a novel type for quite novel purposes' that, 'extends the old symbolic vocabulary beyond its established limits'.[41] All these points can be applied to the development of the ninja myth, although it differs from most invented traditions because of its markedly dynamic nature. Not only has the creative process behind the ninja myth taken place over a period of several hundred years, the invention also seems to be still going on, with the ninja image receiving a further makeover in the run-up to the 2020 Tokyo Olympics.

It is my personal opinion that any tradition that begins to take shape before 1620 and continues to evolve during the present day is worthy of considerable respect as a cultural property, but if the ninja tradition is indeed an invention it is by no means an isolated example, because military societies are particularly vulnerable to the creative approach. Take, for example, the image of the kilted Scottish Highlander, a development discussed in an essay by Hugh Trevor-Roper in Hobsbawm & Ranger's book. The kilt was an invention of an English Quaker from Lancashire called Thomas Rawlinson, who established a furnace near Inverness in 1727 and found that the traditional belted plaid (a large piece of cloth wound round the body) was unsuitable for factory workers. He therefore modified it into the shorter kilt that is still worn today, and after the Battle of Culloden in 1746 the formation of highland regiments in the British Army further developed the image of the Highlander because their members were exempt from the ban on Highland dress introduced by the Disarming Act of 1747. The army therefore 'kept the tartan industry alive' and gave great impetus to the wearing of the kilt.[42] Similarly, the popular image of the pirate derives almost entirely from the American author and illustrator Howard Pyle (1853–1911). Pyle regarded the look of ordinary eighteenth-century sailors as too dull for these romantic figures, so he added elements drawn from the Spanish folk costume of his day – headscarves knotted behind the head, large hooped earrings and wide trailing sashes – to a somewhat mistaken notion of authentic seafaring dress.[43]

Japan is certainly not exempt from this process and a wide-ranging set of Japanese invented traditions was discussed in a fascinating book published in 1998.[44] Elsewhere, in Robert Smith's study of imperial Japanese weddings and funerals in the twentieth century, he concluded that the rituals associated with them were not ancient rites but new ones.[45] As was stressed above, the Tokugawa Period was particularly good at inventing tradition through the rewriting of Japanese history, and shinobi were not the sole beneficiaries of the process. The transformation of the ill-fated Taira clan into tragic heroes and 'loveable losers' is a prime example, and sometimes the invention of tradition was almost an overnight process.[46] The story of the vendetta carried out by the Forty-Seven Rōnin of Ako in 1703 derives almost entirely from an unashamedly fictionalized play that was staged very soon after the incident on which it was based.

To set this juxtaposition of dates in context, the loyalty and self-sacrifice of the Forty-Seven Rōnin was fundamental in establishing the samurai tradition, yet by the time the actual historical incident took place the shinobi/ninja tradition in its earliest form was at least fifty years old. Similarly, the modern understanding of the virtually sacred *bushidō* owes much to *Bushido: The Soul of Japan*, a book first published in 1900 by a Japanese Christian living in America.[47] As *bushidō* is such an important concept for defining the way of the noble samurai, it is ironic indeed to consider that the roots of the ninja myth (perhaps the ultimate anti-*bushidō* concept!) predate it by almost three centuries.

To summarise the above section: the Tokugawa Period's penchant for inventing tradition holds up a huge distorting mirror to anyone seeking to understand the true nature of undercover warfare in pre-modern Japan. The benchmark ninja and his martial art of ninjutsu may be the products of the 1950s and 1960s, but the separate strands that made him date from before the 1650s. The subsequent development of the elements would be long, complex, multi-layered and inconsistent, with different images and different activities included in a contradictory fashion from time to time. These twisted threads were then woven into the ninja of today, giving him a remarkable if convoluted pedigree that even predates much of what we understand as the samurai tradition. I therefore believe that a comparison with the usually unchallenged yet highly questionable 'samurai myth' means that the ninja myth deserves our admiration and fully justifies the academic attention that it is currently receiving from Mie University.

Samurai and ninja

My use of the phrase 'samurai myth' indicates a personal belief that the samurai tradition has itself been subject to considerable exaggeration and commercialisation over the years. There is quite a mythology lurking in the supposedly historical samurai practices of issuing challenges, pledging fidelity and undertaking noble deeds, much of which derives from the romantic *gunkimono* (war tales) that were written from the eleventh century onwards for the entertainment and improvement of their aristocratic readers. In addition, the whole world of the martial arts, with their confusing and competing lineages and multiplicity of styles, was largely the product of the peaceful Tokugawa Period. Schools and traditions (*ryūha*) of hand-to-hand weapon fighting only emerged when personal military skills became less critical, not more. The *bugei* and *bujutsu* (military skills and techniques respectively) that then developed had very little to do with training for warfare during the Sengoku Period, because the campaigns and the armies of the civil wars had been too vast to allow proper individual training to take place. Instead the arts of the sword and the spear became the gentlemanly pastime of the military gentlemen (a literal translation of *bushi* 武士) when wars had ceased, and were taught initially by a maverick group of wandering swordsmen. It was a highly charged spiritual environment in which new weapons could be tried out and old philosophies could be applied to the ideals of personal combat embraced by idle samurai who had read about the deeds of their ancestors.

One very interesting aspect of the ninja myth that is often presented in novels and films is to regard ninja and samurai as totally different types of people. Sometimes the distinction is a simple one involving social class: the samurai are nobles and the ninja are peasants. On other occasions moral values are invoked and the samurai's example of undying loyalty to his lord is compared to the unseemly mercenary ways of the ninja, who is assumed to have fought for anyone who paid him. Further contempt for the ninja is drawn from the observation that in battle the noble samurai squares up to his opponent with a heroic challenge, while the ninja behaves like the thief he really is and attacks secretly in the dark. Even the ninja's useful role as a spy can be contrasted negatively with the fighting samurai, who goes into battle to die selflessly for his lord. The ninja, whose primary goal is the

17

acquisition of intelligence, avoids as much confrontation as possible to stay alive and deliver his message, so there is also no honourable suicide available for a shinobi. Thus in all facets of their existence the anonymous ninja is assumed to provide a dark antithesis of secrecy, mystery, deception and even criminality to the overt, honourable and proud samurai.

There is, however, an alternative to the dark deceiving ninja. This is the positive ninja image noted earlier, which depicts them in movies as underdogs fighting oppression by cruel samurai. These 'good guy ninja' are poor but honest sons of the soil who train themselves in martial arts and resort to trickery and deception because they need guerrilla warfare to survive against large samurai armies. The oppressive samurai then fall victim to their ambushes and guile. This model is a very useful vehicle for displaying the ninja's unrecognised talents and quick-wittedness on the silver screen, and in a variation on the same theme good ninja from a higher social class act as the loyal servants of the Shogun and thwart rebellions by bad samurai and even bad ninja.

All these contrasting extremes involve a considerable manipulation of historical reality, but there is one genuine and very large difference between the samurai and ninja traditions and it concerns their diffusion within Japan. Examples of the samurai tradition are to be found all over the country through links to the former lords of ancient castle towns. In most places there will be a museum, often based inside a restored or rebuilt castle, where arms and armour may be studied and Japan's samurai history explored. The ninja tradition is strangely different, because until its modern commercial exploitation it was almost exclusively concentrated in the two small contiguous areas called Iga 伊賀 and Kōka 甲賀. Iga once meant Iga *kuni* (province) 伊賀国, and when the provinces were abolished in the late nineteenth century Iga merged with neighbouring Ise Province to become part of the new Mie Prefecture 三重県. The former Iga Province (which will be the intended meaning when I use the word Iga) is now divided between the two *shi* of Nabari 名張 and Iga 伊賀, the modern administrative districts that are confusingly translated as 'city', even though a *shi* may often contain much that is rural. Nabari City in the south-west of the old province dates from 1954, and in 2003 its inhabitants voted against a merger with the proposed Iga City that was formed in 2004 by a merger of the former Ueno City and some neighbouring municipalities. Iga City covers the former provincial castle town containing Iga-Ueno Castle and is the nerve centre of the exploitation of the ninja as a tourist attraction.[48] In terms of ninja commercialisation Nabari appears to have surrendered completely to Iga City apart from Akame, a heavily wooded area near the border with Nara Prefecture where the attractive 'Forty-Eight Waterfalls of Akame' offers ninja training based on a very tenuous historical connection.

Iga's northern neighbour provides more of a challenge. The name is pronounced 'Kōka', but until comparatively recently 'Kōga' tended to be used when someone was writing about ninja. The historical Kōka was about four-fifths the size of Iga, but was never a province in its own right, merely a *kōri* or *gun* 郡 (district) of the much larger province of Ōmi 近江, today's Shiga Prefecture 滋賀県. In 2004 the former Kōka-gun was split between Kōka City 甲賀市 and Kōnan City 湖南市. Just to confuse matters there is a place within Kōka City called Kōnan 甲南, which may explain why sketch maps in popular books about ninja sometimes show Kōka only as a tiny strip of land along Iga's northern border, whereas it was much larger. In the pages that follow 'Kōka' will mean the whole of the old Kōka-gun and not just modern Kōka City.

The old provincial border between Iga Province and Kōka District is now the modern prefectural border between Mie and Shiga, which partly explains why the commercial exploitation of the ninja has developed separately in the two areas since it began in the early 1950s. In the 1954 edition of Nagel's *Japan Travel Guide* Iga is mentioned only as the birthplace of the poet Matsuo Bashō,[49] but the city now has a well-developed ninja tourist industry that eclipses Kōka's, as any train-traveller will attest. Kōka Station sports a rather fine bronze ninja statue, but is otherwise fairly nondescript, while the little local train that takes passengers from mainline Iga-Ueno Station has ninja painted all over it and at downtown Iga City Station one is greeted by a dummy ninja at the ticket gate. The traveller can then visit the ninja house and the ninja museum, eat ninja noodles, enjoy martial arts displays and purchase a wide range of souvenirs including ninja toilet paper and the locally brewed ninja beer, which I highly recommend after a hectic ninja-filled day. It is all very good fun, but it is often difficult to distinguish what is being presented as fact and what is acknowledged to be fiction, even inside Iga's famous Ninja Museum.

Such confusion extends much further than misleading displays in glass cases, because the common acceptance of an underlying reality for ninja has meant that is now almost impossible to read an academic text about the Iga or Kōka areas without finding some reference to their most famous sons. For example, Yūki's account of the medieval castles in Kōka bears the subtitle, 'The ninja castle network', and Yokoyama's study of Oda Nobunaga's campaigns in Ise and Iga notes that in the latter area 'the name of the Iga ninja is celebrated'.[50] This also applies to their traditional activities; Yamamoto's recently published study of Japanese military intelligence contains three references to ninja.[51] Outside Japan, Ferejohn and Rosenbluth's *War and State Building in Medieval Japan* begins with a paragraph about ninja who 'existed sometime in the mists of Japanese history'. Its authors clearly take the ninja myth for granted because they regard the ninja of Iga as:

> ...one manifestation of fierce and extensive resistance to encroaching armies in the dying years of medieval Japan. Local farming communities, particularly those in mountain valleys, armed themselves with simple weapons and guerrilla techniques to forestall the trend towards territorial consolidation and centralized taxation.[52]

This is a perfect summary of an important political trend of the times, but to link it to the existence of ninja must not remain unchallenged, and such challenges have indeed been mounted in the past. As Sugiyama reminded everyone in 1974, 'Nowadays ninja are regarded as the stars of Sengoku battles. However, there are very few authentic historical records [about them]'.[53] The overall aim of this book is to trace how this stardom was achieved and to provide some clues as to why it happened.

Chapter 2

The Elusive Ninja

If the ninja myth has any historical validity at all then the evidence for it must lie within reliable contemporary accounts of secret military operations performed during the time of Japan's civil wars. At first sight there appears to be much to choose from, although caution must be exercised over its interpretation because of the previously noted tendency towards exaggeration when most of it was written down during the Tokugawa Period. It is therefore necessary to look back beyond the year 1603 when attempting to establish continuity or to flesh out a reality for the various stereotypes associated with the benchmark ninja, such as his mercenary activity, his precise geographical origin or his espousal of a uniquely Japanese tradition of undercover warfare. The latter point is discussed first.

The ninja's Chinese ancestry
Most popular books view the ninja as being thoroughly Japanese, a product of native ingenuity moulded by local circumstances. There is usually some acknowledgement of a vague Chinese influence dating from the mists of antiquity, but the masked man in black is essentially regarded as the supreme Japanese warrior who belongs to no other culture. Taken to its extreme, the ninja is believed to represent a uniquely Japanese version of undercover warfare and demonstrates a uniquely Japanese expertise in such matters.

The first observation to make on this point is the obvious one that activities conducted in secret on or off the battlefield are common to all martial societies in world history, so it would be surprising indeed to find that Japan, with its long military tradition, was an exception. As for the historical evidence of Japanese uniqueness, one notes that the earliest mention of any spy in Japanese history is of someone who is not Japanese. The reference appears in *Nihongi* in AD 720. The secret agent is a Korean, and the brief entry for the year 601 reads, 'Autumn, 9th month, 8th day. A spy from Silla named Kamata came to Tsushima. He was forthwith delivered up to the Governor, who banished him to Kamitsukenu (Kōzuke Province).[1] The ideographs used to describe Kamata are *ukami hito* 間諜者 ('spyman').[2] The first two characters, *ukami*, can also be read as *kanchō*, the word for spy that would be encountered many times in the future.

The first authentically recorded spying mission by a Japanese person is probably the one that appears in *Shōmonki*, which deals with the suppression of the rebel Taira Masakado and was probably completed shortly after his death in AD 940. The spy was called Koharumaru, and was promised lavish rewards and promotion once the task was completed. He and a companion infiltrated Masakado's headquarters and studied its defensive layout. The second man then returned to their employer, Taira Yoshikane, who launched a night attack, but Masakado's army resisted so tenaciously that forty of the attacking army were killed. During the fighting the spy Koharumaru was exposed and put to death.[3]

Another spying episode occurred at the start of the Gempei War during the first campaign of Minamoto Yoritomo against the Taira in 1180. It is recorded in the highly reliable *Azuma Kagami*. Taira Kiyomori had ordered Yoritomo's capture and execution, and the command to act was given to Taira Kanetaka, the Deputy Governor of Izu Province. Yoritomo planned a pre-emptive strike against Kanetaka's headquarters of Yamagi. The raid was planned for dawn on 8 September, and because it was a difficult target Yoritomo carried out surveillance:

> To offset these difficulties Yoritomo a few days ago ordered Kunimichi, a wandering visitor from the capital who, by an act of fate, had been recommended to him by Morinaga, to make his entry into Kanetaka's domain and to make a sketch of the terrain and the residence. Kunimichi joined Kanetaka at the latter's residence, where he was entertained with song and drink, and in the course of several days he made a complete sketch of the terrain.[4]

It has never been suggested that Minamoto Yoritomo hired Kunimichi because the latter was a specialised shinobi, nor is Yoritomo's inspiration likely to have come from an exclusively Japanese tradition.

Instead the well-educated Yoritomo is far more likely to have been prompted to carry out surveillance from his study of the Chinese classic *The Art of War* by Sun Zi (孫子 otherwise Sun Tzu).[5] Intelligence gathering is fundamental to this classic work from antiquity. Warfare, in the sense of the clash of armies on the battlefield, is viewed as a last resort. A general should only attack when the enemy has been made vulnerable from within by using spies to gather information and infiltrators to create divisions in the enemy ranks. At the point when the enemy has no secrets left and is already defeated in spirit, the wise general launches his attack.

These two elements of spying and causing dissension would be fundamental to the use of undercover warriors in Sengoku Period Japan, and Sun Zi's classic notion of five categories of secret agent would be quoted time and again in the Japanese military manuals of the Tokugawa Period. The character Sun Zi uses for them is *kan* 間, which has the meaning of 'the space between two objects', or 'discord', an obvious reference to the ability of secret agents to cause division between allies. *Inkan*, 'native agents', refers to the employment by one's own side of inhabitants, usually villagers, of an enemy's province or country. As they cannot be properly trained, nor expected to take risks, their usefulness is limited to discovering the enemy's approximate dispositions. *Naikan*, 'inside agents', are the officials of an enemy government whom one's own side takes into its pay. There are many reasons why such people may be tempted to betray their own masters, such as having been passed over for high office. They are the perfect creators of discord, and the information they supply will be of much greater use that that obtained from a mere villager. *Yūkan* literally means 'friendly agents', but the context explains that the understanding is of them as double agents. *Shikan*, 'dead agents', is the chilling title for those of one's own men who are regarded as expendable. They are deliberately given false information and sent on their way. They may well be discovered and put to death, but the misleading information will already have done its work. *Shōkan*, 'living agents', go boldly into enemy territory and return with valuable information. For this purpose men who are pre-eminent in intelligence are selected. They have the ability to withstand great hardship and have the social skills to gain access to those of the enemy who are closest to the seat of power. These are the classic spies envisaged by the ninja myth.[6]

There are enough references to *The Art of War* and other Chinese military classics in Japanese historical accounts to demonstrate that the samurai aristocracy were very familiar with their contents, including the material on spying and undercover operations. The early *gunkimono* also contain several references to Sun Zi's recommendations. The samurai hero Minamoto Yoshiie (1041–1108) is said to have thwarted an ambush by noticing birds rising in a disturbed fashion from a wood, a clue he had learned from *The Art of War*.[7] Minamoto Yoritomo's brother Yoshitsune (1159–1189) is supposed to have seduced a girl just to get the chance to read a copy of a Chinese military manual owned by her father.[8] Turning to more historical individuals, Mōri Motonari (1497–1571), the *daimyō* of Aki Province, was a leader who certainly knew his Sun Zi and put the teachings successfully into operation,[9] while Takeda Shingen (1521–1573) was so impressed by *The Art of War* that he had a quotation from it emblazoned on his battle standard. Japan's ancient corpus of military techniques involving spying is therefore more likely to have had Chinese origins than to be uniquely Japanese. The crucial question for the ninja myth is therefore not where the Japanese obtained the idea from, but how they exploited it.

The shinobi as members of a hereditary élite

One important element within the ninja myth states that the Japanese exploited Sun Zi's original ideas through specialisation, so that secret warfare in Japan was carried out only by a highly skilled hereditary corps of élite warriors called shinobi. Ninjutsu, however it might be defined, is therefore regarded as the exclusive preserve of these 'special forces'. The authentic historical accounts that follow are indeed the stuff of special force operations, and the idea that they were performed by a force of medieval commandos allows modern authors to slide effortlessly from talking about ninja to describing the activities of Green Berets or Navy Seals. These are tempting analogies, but the élite model of the ninja's origins is complicated by the fact that the members of Japan's samurai class were precisely that: highly specialized, hereditary and élite. The historical shinobi would therefore have to be better than they were, in other words the 'super-samurai' beloved of modern ninja movies. In that case no ordinary samurai, no matter how good he may have been at the martial arts, could perform these techniques of castle-entry, battle-disruption or intelligence gathering.

The contemporary evidence for shinobi as either 'super-samurai' or dedicated special forces within a *daimyō*'s army is quite difficult to interpret. Undercover operations are never hard to locate in Japanese history and a few, like the *Taikei ki* account quoted earlier, imply élite operations, yet only some of these exciting stories identify specialised forces from Iga or anywhere else, and even fewer of them would feed into the ninja myth. Examples of ones that did not include an incident in 1510 when the Sō family of Tsushima were involved in the 'Three Ports Incident' of 1510. This was a series of armed disturbances by Japanese residents within their specially designated three trading ports on the southern coast of Korea. The trouble-makers were supported by the Sō family of Tsushima, and the Korean account of the bloody affair refers to secret forces called *shinomi* 時老未 from Hakata in Japan infiltrating Korean positions.[10] In Western Japan *Chūgoku Chiran ki* describes the Mōri family of the Inland Sea area using *shinobi no tsuwamono* 忍びの兵 (secret soldiers) against their rivals the Amako, but they are not identified as special forces and they are certainly not Iga men.[11] Another is the occasion when Ikeda Nobuteru captured Inuyama Castle on behalf of Toyotomi Hideyoshi in 1584. He and his troops crossed the Kiso River at the dead of night using small boats and entered the castle through its water gate. The operation is simply recorded as a surprise attack. There is no suggestion of the raid being carried out by specialist shinobi forces within the army. It is just a clever secret operation performed by some of Ikeda's bravest samurai.[12]

There are also several examples in the early and possibly exaggerated Tokugawa Period war chronicles. We read in *Ōu Eikei Gunki* of 1698 a splendid story of a man who is skilled in shinobi activities, but is not a member of any hereditary élite other than being a samurai. The incident occurs during the siege of the castle of Hataya in 1600, where he performs a clever piece of psychological warfare:

> Someone inside Hataya castle with renowned *shinobi* skills 忍びの上手 (*shinobi no jōzu*) that night entered the enemy camp secretly (*shinobi iri*). He took a *ban sashimono* (company flag) from Naoe Kanetsugu and a *fukinuki* (streamer) from the barracks of Kurogane Magoza'emon, returned and hung them from a high point on the front gate of the castle.[13]

The word shinobi appears twice in the above account to describe the man's skills and activity, but we must resist the suggestion that the word means he is a superman or that he belonged to a specialized hereditary élite. Like Ikeda's warriors in 1584, he was just 'someone' who had these particular talents and used them to his side's advantage.[14] Other Tokugawa Period examples include *Mogami Yoshimitsu Monogatari* of

1698 where there are references to infiltration including 'secretly entering and killing' and 'secretly entering at dawn'.[15] In *Date Hikan* of 1770 the Date family of Sendai are credited with using their own variety of secret warriors during Date Masamune's battles of the Sengoku Period. They are very like special forces and are referred to as the *kuro habaki gumi* 黒脛巾組 (the black *habaki* group), *habaki* being the leggings worn below the knee.[16] We must allow for a possible exaggeration because of the late date, but this would appear to be the only authentic reference to specialists operating undercover while wearing black.

Shinobi and thieves

There is, however, a different way of looking at the shinobi's élite status, because many movie plots suggest that much of the undercover work was done by people who came from the opposite end of the social spectrum. If this model has any basis in fact then these men were a lower class élite who acquired their shinobi expertise simply because of a need to survive. They may even have been a criminal élite, as is suggested by the numerous occasions when they are identified using character compounds like *settō* 窃盗, which normally means thieves. We have already noted that in *Arayama kassen ki* of about 1620 a raid is conducted by a 'band of robbers from Iga'. In 1554 *Goseibai Shikimoku*, the legal code first set up in 1232 and maintained for many centuries, gives a useful list of words for thieves, many of which reappear later in the context of shinobi operations. So along with *settō* we find *nusubito* 盗人 (thieves), together with *konusu* 小盗 (in modern parlance 'petty thieves') and *gōtō* 強盗 (literally 'strong thieves', now meaning burglar). All these words have appeared in accounts of secret operations from time to time. The word *gōtō* is in the *Moromori ki* account cited earlier of the night attack on the Yūki family's castle in 1360.[17] The Yūki family records also use the word yowaza 夜業 ('night tricksters') for their spies employed in Shimōsa Province in 1584.[18] *Gōtō* also appears in *Buke Myōmokushō* of 1806, which says that shinobi were drawn from among the *gōtō*.[19]

A very good example of the casual association of shinobi with thieves appears in the opening sentence of the sixth chapter of the 1653 military manual *Gunpo Jiyoshū*: 'If a *daimyō* does not have *settō* serving under him then no matter how good he is he will know nothing of his enemy's dispositions'. In his presentation of this passage the modern military writer Sasama Yoshihiko glosses the word *settō* as shinobi.[20] The same characters *settō* are also found in *Yoshimori Hyakushu*, a series of poems

26

on secretive military themes incorporated into the above *Gunpo Jiyoshū*.[21] Similar words are even found in the so-called 'ninjutsu manuals' of the seventeenth century. *Mansenshūkai* of 1676 refers to *yatō* 夜 盗 (night thieves).[22] The idea is also very common in modern literature and films. Ishikawa Goemon, an outlaw whose life has acquired the veneer of a Japanese Robin Hood, has often been portrayed in films as a ninja, the better to show off his cunning.

So who were these people who are so easily and regularly identified as thieves? A clue appears during the fourteenth century in *Mineaiki*, which was compiled in 1348 by a priest from Harima Province (part of modern Hyōgo Prefecture):

> Various kinds of disturbing events occurred around the eras of Shōan and Kengen (1299–1303), with rebellions, coastal piracy, raids, robbery, mountain banditry, pillaging and so on happening all over the place. They disguised themselves in an unusual way by wearing yellowish brown clothes and a *roppōgasa* hat like a woman's instead of an *eboshi* (cap or hat) and not showing their faces. Individuals who congregated in groups of between ten and twenty men wore swords that had no ornamentation, with rough quivers on their backs and bamboo poles for spears and neither helmet nor armour. They withdrew to castles and took on their enemies there, or they won over an enemy but then betrayed him, committing themselves to nothing. They were fond of gaming and gambling, and they behaved like *shinobi konusu* 忍び小盗 (sneak thieves).[23]

The contemporary name usually given to these groups of marauders was *akutō* 悪党.[24] Nowadays the word simply means bandits, but *akutō* had a much wider meaning during the fourteenth century, so Karl Friday sensibly translates the word as 'evil gangs'.[25] In 1318 steps were taken to eradicate them, resulting in the deaths of a few *akutō* members and the destruction of twenty of their forts, although most of the gang members remained at large.[26] Yet even though *Mineaiki* dismisses them as 'sneak thieves' in the above section, by about twenty years later their image had changed. Now the author describes them as forming bands of fifty to one hundred horsemen, led by mounted warriors wearing fine armour and carrying weapons ornamented with gold and silver. They are still feared because 'they chased people out of their own homes, raided their properties and stole their crops'. However their 'gangs' now consist of members who have pledged loyalty to a leader, so it would appear that the *akutō* are in reality no more than local warriors asserting their independence from a remote central government.[27] The conclusion must therefore be that higher class commentators lumped them in with thieves, pirates and bandits, and the same thing would happen two centuries later. The Yūki family records, for example, use the overall expression *akutō* to identify the spies who served them in 1584.[28]

This use of contemptuous expressions for the lower orders of society is a practice we will encounter later in the context of the *ikki*, or leagues, who challenged *daimyō* rule during the sixteenth century, and in fact Iga Province provides excellent examples of both usages stretching over several centuries. In the words of Elisonas, 'Iga, a small but intractable province… was a notorious lair of "bands of evildoers" (*akutō*), that is, independent-minded country samurai…'[29] The fourteenth century *akutō* were similarly feared for their raids on important communications routes and equally despised for their presumption. They did, however, provide a useful pool of recruitment in times of war, along with other named groupings of socially inferior warriors such as those called *nobushi* (野武士 or 野伏, 'field warriors', sometimes translated as bandits) who earned an additional level of contempt when they sold armour and weapons plundered from dead samurai. Under the general term of *akutō* small landowners, pirates and other despised social groups took part in the guerrilla campaigns of Kusunoki Masashige in the mountains of Yoshino and contributed towards the idea of him as a leader of ninja.[30]

The specifically identified 'Iga Province evildoers' crop up in accounts of *akutō* activity in the southern half of Iga (modern Nabari City) from the eleventh century onwards. It was noted above that Nabari does not exploit its ninja connection in any way comparable to Iga City, even though the social and historical background cannot have been very dissimilar. The *akutō* activities for Nabari relate to a place called Kuroda, an estate owned by the Tōdaiji temple of Nara.[31] There were repeated instances of conflict in Kuroda over a failure to pay taxes to the provincial governor by residents who had taken on the mantle of the tax-exempt temple. Other inhabitants then tried to set themselves up as minor feudal lords, which brought them into conflict with the formerly sympathetic Tōdaiji. From the late thirteenth century onwards the term *akutō* began to be used to describe the assailants and in 1280 Tōdaiji denounced three leaders in particular in those terms. The evildoers were called Ōe Kiyosada, Ōe Kiyonao and Hattori Yasunao, and all three were named in petitions to the government about their criminal behaviour:

> Not only is it clear that the residents of our Kuroda estate in Iga province, Ōe Kiyosada, Yasuna, and others have committed such evil acts (*akugyō* 悪行) as brigandage, night raids, robbery, arson and murder, they have also piled crime upon crime against the proprietor by blocking roads, contracting barricades and fortifications, and making great disorder, crimes for which they should not go unpunished.[32]

The shinobi as a mercenary

This idea of shinobi as lower class warriors with a shady background also provides a clue to the popular understanding of them as mercenaries. The notion that men from Iga and Kōka in particular sold their services to others is repeated time and time again in films and in the popular literature on the subject, but the concept is virtually meaningless if we cling to the idea of a shinobi as a super-samurai. A mercenary on the strict European model is a soldier who comes from outside the society for which he fights, who is not part of its regular forces and who is motivated primarily by the desire for private gain. The prime examples were the notorious mercenary bands of Europe who would fight for one Italian city-state one month and fight against it the next.[33] This professional model of *condottierri* or 'soldiers of fortune' was totally absent from the Japanese scene, although some Japanese warriors did serve as mercenaries overseas. Between 1593 and 1688 Japanese fighting men, most of whom were exiles and many of whom had experience of piracy, were in the service of the kings of Siam and Cambodia, the Spanish colonists in the Philippines and the Dutch East India Company. They were nearly all drawn from the communities of expatriates who lived in enclaves in places like Manila, Macao and Malacca, although for a brief period of time the Dutch East India Company hired samurai from Japan itself and shipped them overseas to fight.

Mercenary warfare like this is totally unrecognisable within Japan itself until we shift the focus on to the lower classes of Japanese society. At this level there was a genuine use of mercenaries in Japanese warfare. Yamada uses the word *yōhei* 傭兵 (mercenary) to describe the recruitment of *akutō* by the various warring factions during the

Nanbokuchō Wars,[34] and Fujiki's study of infantry on the battlefields of the Sengoku Period even has as its subtitle 'medieval mercenaries and slavers'.[35] Mountain bandits and pirates provided a useful labour force in this regard, but it is very noticeable that the mercenaries did not behave like conventional samurai. They were not hired to carry out cavalry charges or fight honourable sword encounters. They were used only for sneak attacks, night raids and the creation of havoc within an enemy encampment.

These irregular troops were identified using the words noted earlier like *kusa*, *yowaza* and shinobi, but in northern Japan in particular we also encounter the important names *suppa* 透波 and *rappa* 乱波.[36] The two names combine the characters for penetration and disorder respectively with that for a wave. The role of the *suppa* was largely that of a spy. They are associated in particular with the Takeda family and we read that Takeda Shingen had about 500 *suppa* from Shinano 'on his books'. They sneaked into castles on moonless nights and on one memorable occasion the enemy guards threw stones and pine torches down from the walls to try and detect them.[37]

Rappa are particularly associated with the Takeda's contemporaries, the Hōjō family of the Kantō, for whom *rappa* were one of three types of undercover warriors who performed distinctive tasks on the *daimyō*'s behalf. At the highest level were the Hōjō's mobile mounted 'warrior scouts' (*monomi no musha* 物見の武者) of samurai rank, who 'were highly trained horsemen, and exclusively persons of great merit', according to *Hōjō Godai ki*, which was compiled between 1641 and 1659.[38] Then came the men of ashigaru (foot soldier) status who spied on the enemy while concealed in the long grass, hence their name *kusa*. The word shinobi was also used for them, thus providing a further instance of the expression in its noun form to identify men who served within the ranks of a *daimyō*'s army, but the *rappa* were quite different both in role and origin. Introducing the subject in *Hōjō Godai ki*, author Miura Jōshin tells us:

> In olden times, as we have seen, chaos reigned in the various provinces of the Kantō, so we always had to keep our bows and arrows at the ready. Now there were in those days rogues called *rappa*. These individuals were thieves, but were not just thieves because they were both wicked and cunning at the same time. In the old writings they are called *rappa*, but the way the word is written is not always clear. These people were granted a stipend in the provincial *daimyō*'s army. Whatever the reason for what they were

30

called, all these *rappa* skilfully sought out the bandits in their own provinces, hunted them down, and cut off their heads. They also sneaked into (*shinobi iri*) neighbouring provinces to carry out mountain banditry, piracy, night attacks and abductions of their own. The Kantō *rappa* were intelligent and devised plots and plans unachievable by ordinary people. In terms of their wisdom they were almost gods or buddhas.[39]

The reference to them receiving a stipend or allowance (*fuchi* 扶持) confirms their mercenary status, but it also indicates their partial integration into the Hōjō army. Jōshin's gushing account of them in action is worth quoting in its entirety:

[Hōjō] Ujinao had among his command two hundred *rappa* who received stipends, one of whom was a wicked man. His name was Kazama. He was regarded as a top villain with superlative skills. Under Kazama were four officers. One was a mountain bandit, one a pirate, and the other two were *gōsetsu* (robbers). The bandits were familiar with the terrain of the mountains and the rivers, and could therefore sneak into enemy positions. These four men primarily led night attacks. Their unit of 200 men was divided into four sections, and went out whatever the weather, on rainy nights or dry nights, still nights or windy nights.

Every night or so they crossed the great Kisegawa and entered secretly into Katsuyori's camp. They captured people alive and cut through the ropes tethering the horses, which they rode bareback, plundering and raiding to ridiculous lengths. In their night attacks, moreover, they set fire to things, and raised the allies' battle cries to make them think they were friends. All the camp was in uproar and shock. Armour belonging to one was fought over by two or three others with fierce arguments. Panicking to get out they were led astray to front and rear. Thinking them enemies they turned against their friends, killing each other, scattering fires, and putting all their plans into disorder, completely confused. When dawn broke they examined the heads of the slain, and discovered that in the fighting low-ranking soldiers had taken the heads of their lords, and children had taken the heads of fathers.[40]

Jōshin also provides a bizarre description of their celebrated leader Kazama Kotarō as

31

a giant who cannot be concealed among his two hundred men. He is 7 *shaku* 2 *sun* in height, with roughly hewn sinews in his arms and legs. His intelligence raises him above the common herd. His eyes appear to be upside down, he has black whiskers, and his mouth is particularly wide at the sides. Four of his fangs stick out. His head resembles Fukurokuju and his nose sticks up.[41]

The Hōjō documents draw a clear distinction between the mercenary *rappa*, whose primary aim was to create havoc, and the shinobi/*kusa* within the Hōjō's own ashigaru, who had an intelligence-gathering role. The latter would hide in the long grass for hours on end, often returning the following day to report, and *Hōjō Godai ki* describes how both the Hōjō and their rivals the Satake would send *kusa* to spy on the enemy lines.[42] *Kusa* used by the Satake once tried to intercept the Hōjō's mounted scouts as they returned from a reconnaissance mission, springing up to attack them. 'They rose up like bees to surround the two mounted men and catch them like fish in a net. They made to take the horse of San'emonnojō, but in spite of being in enemy territory he turned his horse to the north and whipped it up'. As a result of their horsemanship skills both *monomi no musha* 'escaped from the jaws of the crocodile'.[43]

Variations on the word *kusa* and the practices from which it derived are found throughout Japan. In the diary *Yatsushiro nikki* of the Sagara *daimyō* of southern Higo Province (modern Yatsushiro City, Kumamoto Prefecture) there are references to night attacks being carried out on various occasions between 1540 and 1564 by men called *fusekusa* 伏草 ('those who hide in the grass'), and sometimes these marauders were recruited from local mountain bandits.[44] Both the Shimazu and the Date families from the geographical extremes of Japan used men referred to as *shinobi no kusa* 忍びの草.[45] During an operation by the Ashina family in 1588 in what is now modern Fukushima Prefecture, infiltrators were sent from Tamanoi to Takatama to 'lie down in' and 'spring up from' the grasslands in the guise of *kusa*.[46]

All these accounts clearly suggest that in addition to the tasks carried out by a *daimyō*'s own soldiers, there was a considerable amount of mercenary activity whereby local hoodlums and ne'er-do-wells received payment for carrying out high-risk operations. These activities even included using men as human booby traps when an army made a tactical withdrawal. The soldiers left behind in the long grass in that situation would spring up and hamstring the pursuing enemy horses.

This practice suggests that some irregulars were recruited because they were expendable, an attitude that adds a further dimension to how despised these ruffians were and detracts still further from any notion of them as super-samurai. The highly valued and respected Kantō *rappa*, who appear to have been hired 'on extended contracts' may be an exception, but at the furthest extreme there were occasions when even a willingness to fight was lacking and the irregulars had to be forced into action. According to the twenty-second chapter of *Kōyō Gunkan*, when Takeda Shingen was fighting on the border between his home province of Kai and neighbouring Shinano, he assembled seventy men from Shinano to act as *suppa*. Out of them he chose the thirty most highly skilled practitioners, but before sending them off to spy in their own province he took hostage their wives and children as a guarantee that they would return with the information.[47]

The other use to which mercenaries could be put was the sordid one of kidnapping and abducting people of their own kind to use as forced labourers. This brutal process is referred to as 'slave harvesting' (奴隷狩り *doreigari*). In contrast to medieval Europe, where a knight might be seized and held to ransom, the victims in Japan were poor farmers and ransom only took place when so many slaves had been gathered that the captors could no longer cope with them:

> Between Sagami and Awa there is but a short stretch of sea and both sides have ships. The fighting never stops. They sometimes come across in small boats during the night to raid the coastal settlements. Sometimes they have thirty ships and set fire to villages on the shore, seizing women and children and putting to sea while it is still dark. The inhabitants of Shimazaki have come to a private understanding with the enemy and pay them tribute in rice so they can live in peace at night time. They are in secret contact with the enemy and buy the hostages back.[48]

The chaos of the Sengoku Period allowed such depredations to flourish. Bands of irregulars joined in with the regular foot soldiers in a *daimyō's* army to carry out these acts of *randori* 乱捕り (indiscriminate pillaging and enslaving) and it is more than likely that most campaigns saw violence of this sort. *Kōyō Gunkan* relates just such an episode after the first battle of Kawanakajima in 1553 when the Takeda raided deep into Uesugi territory 'abducting people from Echigo Province and bringing them back here to serve us'.[49] Fujiki devotes several pages of

Zōhyōtachi no senjō to examples taken from diaries relating to warfare in southern Kyushu. 'More than fifty killed and an unknown number of men, women, cattle and horses taken' (1546); 'two hundred and thirty-six heads taken, and many captured alive' (1549); 'two of the enemy killed, besides these fifteen or sixteen children and old people taken' (1555).[50] As to who was responsible for such depredations, the records of the Sagara *daimyō* demonstrate clearly that the abductions were carried out by *fusekusa* hired as mercenaries from the local pirates and bandits. '*Fusekusa* from Hishikaru killed two men and captured two' (1557); '*Fusekusa* landed from the sea, killed thirty men and captured eight' (1559); 'In the night attack from Ikeura and Sashiki one man was killed and two were taken' (1561). As Fujiki sums up so well, innocent people who had just gone to sea to catch fish or up into the mountains to gather firewood were captured and carried off. This was total warfare in its nastiest sense, and the irregulars who carried it out were the same men who also took part in highly dangerous night-time castle attacks under the name of shinobi, creating thereby a tradition that would leave their true nature far behind. Theirs, and not any noble samurai's, was the authentic world of *shinobi no kōsaku* 忍びの工作 (secret operations).[51]

The above accounts suggest that an élite explanation for shinobi does not necessarily imply a social élite, and may even imply a criminal élite, but this model has never been a popular conclusion and adverse reactions to it are noticeable from the Tokugawa Period onwards. In his *Iga Kyūkō* of 1699 the highly biased Kikuoka Jōgen provides an alternative explanation for the strong association between the heroic shinobi of Iga Province and common thieves. He first notes that, 'In ancient times the country samurai of Iga were adept at the principles of shinobi', although because he writes shinobi using the characters for thief, his phrase could be translated as 'had a penchant for theft'.[52] Elsewhere he uses *settō* 窃盗 for shinobi 'as in the word *nusubito* 盗人 for someone who enters and steals things', but he explains then that because persons unrelated to his heroes performed similar actions in their course of their criminal activity the ideograph for thief was affixed to them all.[53] A similar point has recently been made by Kawakami Jinichi. He acknowledges that the word shinobi can be understood as 'thief' because of its association with secrecy and concealment, but because the role of the shinobi included entering an enemy's territory to pillage and create havoc they were hardly likely to acquire a positive image and the label of thief would be an easy one to apply.[54]

Whatever the truth may be, the pre-Tokugawa accounts certainly point to the conclusion that warriors who operated in a shinobi manner came from the widest possible spectrum of social class origins. The periodic dismissal of them as evil gangs of thieves adds a further dimension to the notion of a clear division between noble samurai and ignoble shinobi. One of the key notions lying behind the ninja myth that is regularly presented in modern films, the existence of special forces manned by a super-samurai élite, is therefore highly questionable, while the idea of them as lower-class warriors who possessed criminal skills may not be far from the truth. In other words, if a castle had to be captured or information was needed, instead of using skilled samurai from among his own men or even hiring expensive mercenaries from Iga Province, a *daimyō* simply called upon the group who could best demonstrate the skills of entering places in secret. He sought the services of his local burglars.

So is the truth behind the ninja myth to be found among wretched, hungry, desperate outcasts from the criminal element in Japanese society to whom the emerging warrior class as a whole might often be enemies? The activities of the *rappa* have certainly fed the ninja myth and men like them are plausible models for the ninja's origins. The staunch, anti-establishment, lower-class brigand image also has a certain appeal, as numerous movie plots attest. Even the careless and uncritical handbook that accompanied the 2016 children's ninja exhibition in Tokyo included *akutō* in its list of the ninja's honourable antecedents.[55] Yet other ninja enthusiasts object most strongly to the idea of ninja as criminals, arguing that key historical accounts prove that the shinobi who were recruited by *daimyō* were neither ruffians nor thieves, but came from a closely defined area of Japan called Iga and Kōka where undercover skills were far from being despised. Instead they commanded a high price that was negotiated by a class of 'super-ninja' called *jōnin*. This is a vital part of the ninja myth, and even though I have already suggested that Iga/Kōka did not have a monopoly on secret warfare, the important thing is that for many centuries they have claimed to have had one. The names crop up time and time again in accounts of undercover operations, so it is to the history of these two places that we must now turn.

Chapter 3

Iga and Kōka

One of the best known elements in the myth of the ninja concerns their links to the areas of Iga and Kōka, because whatever may be claimed about the shinobi's social origins, the ninja myth firmly maintains that their geographical origins can be pinpointed precisely. This may sound highly unlikely, but it forms the basis for most of the modern commercial exploitation of the ninja and is also the feature maintained most fiercely by ninja enthusiasts. It is, however, the most difficult one to accept, because it defies all common sense. During the Sengoku Period fighting was carried out from Okinawa to the fringes of Hokkaido, yet we are required to believe that this tiny area of Japan produced a caste of warriors so skilled that their talents were widely exported. Every book on ninja homes in on the connection to the evident delight of the local tourist industry, although in all fairness it must be pointed out that institutions like the Iga-ryū Ninja Museum in Iga City do offer a grudging recognition that secret warfare was carried out elsewhere in Japan.

This is an important point to consider, because there is a very big difference between claiming exclusivity for something and claiming expertise in it, but even if the Iga Museum holds back from a claim of absolute exclusivity the point is always made most firmly that Iga and Kōka specialised in ninjutsu, that the very best ninja came only from there and that any other traditions were developed by emigrants from Iga or Kōka. These watered-down claims may look as ridiculous as the idea of exclusivity, but they must not be dismissed out of hand because to specialise is very Japanese. An expert concentration on skills like horsemanship or swimming is a feature of many of the traditional schools of martial arts, so undercover operations need not be any different. There could also very easily have been a persistent local tradition of excellence in secret warfare. Many areas of modern Japan claim hereditary expertise in things like growing giant radishes or indigo dyeing, so there is no absolute reason why the areas of Iga and Kōka should not be able to claim a particular speciality in spying that was passed on from father to son. The principle of inheritance has always been a strong element within many Japanese traditions and the greatest kabuki actors can look back upon long genealogies within their exclusively theatrical families. The casual statement that, 'the ninja came from Iga and Kōka' may therefore indicate a genuine tradition of expertise, although we must guard against any tendency to think that the arts of the shinobi were the absolute birthright of every inhabitant, even if the movies suggest that they were. In several historical accounts the skills in secret operations often claimed carelessly for all Iga and Kōka men are in fact identified in only a few hand-picked individuals. *Iran ki*'s tale of the defence of Kashiwara Castle in 1581 by heroic local samurai refers to 'ten Iga-mono who had mastered shinobi no jutsu'.[1] When Takatori Castle was infiltrated in 1600 by a Kōka-mono (a man from Kōka) the account in *Mikawa Gofudo ki* describes him as 'one who had shinobi skills' and was hand-picked from the Kōka force because of it.[2] Not every child in Iga was born with a *shūriken* in his knapsack.

It is, however, important to note that the splicing together of Iga and Kōka because of a supposedly shared expertise is far from being a whim of the modern ninja tourist industry. Long before the inflated Tokugawa Period accounts were written the two areas were being linked within authentic historical narratives, and it is not difficult to discover why this happened. Kōka-gun fitted neatly along Iga's northern border where communications were unimpeded by high mountains. Their common attitude towards independence of action also made them natural allies against powerful aggressors, although I believe that only one author has gone so far as to describe the situation using the term *Kōi ikkoku* 甲 伊一国 (the single province of Kō[ka] and I[ga]).[3]

In fact the circumstances surrounding their history meant that the two areas were never one single unit. An important border has always run between Iga and Kōka, and even if that meant little to their inhabitants at the time of the Ōnin War, it meant a great deal to their superiors in Kyoto. Within a century the focus of power slipped from the Shogun to the *daimyō* and there were then even more occasions when it was natural for the two neighbours to join forces, but Ōmi Province consisted of much more than just little Kōka and the pressures brought to bear upon Kōka usually had no reference to Iga. The two areas may have cooperated when they could, but very different political and military influences meant that they developed very differently and then gradually split apart. During Nobunaga's campaign against Iga in 1581 the men of Kōka actually fought against Iga, and they only really came back together after each of them had separately rendered service to the future Shogun Tokugawa Ieyasu in 1582 and 1600 respectively. As their rewards they became two units within the hereditary palace guards of Edo Castle, which meant that men from Iga and Kōka would serve side by side for many more years to come.

Iga Province in Japanese history

The area formerly known as Iga Province, with which we will begin, was defined by rivers and defended by mountains. The mighty Kizugawa arises in its centre and flows down into the Ao Valley where it turns northwards. It then makes a westerly turn near Iga City to flow on towards Osaka Bay. To the north of Iga City it is joined by the westward flowing Tsugegawa, beyond which are some modestly sized mountains and an area of flatland that communicates directly with the historical Kōka District. According to the ninja myth it was within this small, isolated and landlocked area that the whole idea of ninja began, and if that outrageous notion has any basis in fact then there should be considerable evidence of secret operations by Iga warriors that set the province apart from other less favoured lands.

At first sight the supporting documents appear to be plentiful and often describe the actions in exciting detail. However, most of them were composed many decades later and have therefore been open to distortion, so even though they describe events that almost certainly occurred, their information about the nature of the participants has to be seen through the distorting mirror of the Tokugawa Period and cannot be taken at face value. The firm evidence to support the model of an Iga specialism in undercover warfare must therefore be sought in pre-Tokugawa accounts, but immediately a problem is encountered because Iga Province was completely devastated by Oda Nobunaga's invasion of 1581. Almost no local historical records survived the onslaught, so any claims about the Iga ninja having their origins in authentic operations are dependent upon accounts of them written in places other than Iga, and in fact only five authentic contemporary accounts of secret operations by Iga warriors prior to 1581 have survived. All except one are presented here for the first time in an

English translation. It may not be much on which to base so grand an edifice, but each has something valuable to contribute.

The first four accounts are to be found within letters and diaries composed at local temples, while the fifth, which has only recently come to light, is a handwritten manuscript concerning an operation within Iga itself. That one uses shinobi as a noun; the others all display the use of shinobi in its adverbial form to describe attacks by a military force known simply using a conventional term as the Iga-shū 伊賀衆 (the Iga military unit). The accounts concern an assault on Kasagi 笠置 Castle in Yamashiro Province (modern Kyoto Prefecture) in 1541 and three other castle raids in Yamato Province (modern Nara Prefecture). These are Takada 高田 in 1556, Tōichi 十市 in 1560 and Sakaibe 阪合部 in 1580. The sites are located respectively in Takada City, Kashihara City and Gojō City.

The 1541 raid on Kasagi Castle is the only event among the five which is at all well known. It has been widely quoted in books about ninja for many years because the operation appears to provide everything that one could wish for in establishing the historical authenticity of the Iga shinobi tradition. This is because the account is a factual record made by a dispassionate observer writing long before the ninja myth was established with no Tokugawa Period embellishment. The word shinobi is present and men from Iga and Kōka take part, the latter being included within the Iga-shū. The operation is also a consummate exercise in infiltration, which takes place away from Iga Province, suggesting that the Iga men are acting as mercenaries. As a consequence much has been extrapolated from its few sentences to make the reference into the most important proof text for three crucial points: that the élite shinobi no mono came from Iga and Kōka, that they had unique skills and that they exploited them on a mercenary basis.

The incident is covered in two separate entries in *Tamon'In nikki*, a diary kept by Abbot Eishun of Tamon'In, a sub-temple of the Kōfukuji in Nara. The operation against Kasagi Castle was carried out by Tsutsui Junshō (1523–1550), and the background is that of the complicated rivalry between powerful *daimyō* based not far from Kyoto who fought one another for control of the Shogun. Kasagi was under the command of a vassal of Kizawa Nagamasa, and Tsutsui Junshō's siege cannot have been going well, because his ally Hosokawa Harumoto petitioned the Shogun to send a letter to the provincial governor of Iga ordering him to help in the attack. The letter was sent on the 18th day of the 11th month of Tembun 10 (15 December 1541). The priest's diary entry for

39

eight days later (23 December 1541) reveals that men from Iga responded to the call and launched a surprise attack, setting fire to the place:

> This morning, the Iga-shū entered Kasagi Castle in secret (*shinobi itte*) and set fire to a few of the priests' quarters and so on. They also burned down outbuildings in various places within the third bailey and are even said to have seized the first bailey and the second bailey. According to Ukon, the castle commander on the Kizawa side, they were from Kōka in Ōmi and numbered between 70 and 80 men. Maitreya Peak,[4] the place on which [the castle] stood was a mountain with no water at all, so it got to the point when it would inevitably fall. A few men from Sugawa and Yagyū deceitfully changed sides to join Tsutsui Junshō's force who were stationed slightly to the rear.[5]

Two days later in *Tamon'In nikki* we also read:

> As for the battle taking place at Kasagi, the Kizawa side sallied out from the castle in two groups and were slain because they were exhausted, then all the Iga men dispersed, which was some relief although thirty men were dead.[6]

The second incident involving Iga occurred at Takada in 1556. It is noted in *Kyōroku Temmon no ki*, a contemporary work also associated with Kōfukuji. Once again the Iga men are fighting for the Tsutsui:

> On the 12th day of the 12th month of Kōji 1 (24 January 1556) eleven men from the Iga-shū on the Tsutsui side attacked Takada Castle in Yamato. As the Shogun had been delayed, the men inside the castle were all killed, and not only was the castle destroyed by fire but also the temple halls of the Jōkōji that was built nearby.[7]

The attack can be precisely located because the site of Takada Castle is now marked by a monument within the grounds of Takada High School. Next to it is the rebuilt Jōkōji. The third incident happened at Tōichi Castle in 1560 and is also recorded in *Kyōroku Temmon no ki*. One of the most interesting features of this account is the name of the leader of the Iga contingent. He is Shimotsuge no Kizaru, who would later be named in *Mansenshūkai*, the so-called 'ninja bible' of 1676, as an influential leader in Iga. The Tsutsui family are again mentioned, but in the person of Junshō's son Junkei (1549–1584), then called Okamisama and aged eleven:

> ...on the night of the 19th day of the same 3rd month the Iga-shū under the command of Kizaru raided the *yamashiro* of Tōichi [under] Lord Hashio Shōjirō [Tōkatsu]. It is said that the general's residence was captured, so he took the road to Toyoda Castle and his retainer Dōruku. Ueda was killed, Okamisama was absent in Momo-o, four men were killed there.[8]

The fourth account is a report of an attack on Sakaibe Castle in 1580. It takes the form of an enthusiastic letter written to one Futami Mitsuzōin by Ichirobō, a priest of the Kongōbuji on the holy mountain of Kōyasan. This time the victim is an ally of Oda Nobunaga called Sakaibe Hyōbudaiyū, whose castle bears his name. There is no date given for the operation, although the timing of the letter as Tenshō 8, 8th month, 4th day (12 September 1580) shows that it occurred between Oda Nobukatsu's unsuccessful invasion of Iga in 1579 and the destruction of the province in 1581. The people of Iga were obviously not solely occupied with defensive measures at that time; rather they were taking the fight to the enemy:

> In secret and in the middle of the night the Iga-shū entered Sakaibe Hyōbudaiyū's castle in Sakaibe in the Uga District of Yamato Province; they crossed the wet moat from the south and were the first to arrive at each entrance. To go inside the castle was an action without parallel.[9]

The fifth account is completely different and consists of a number of references within a handwritten manuscript called *Amagoisan rōjō okite kaki* 雨乞山籠城掟書 (Written regulations for the siege of Amagoisan). It was preserved within the archives of the ancestors of Kawakami Jinichi, a man who has devoted much time and energy to locating contemporary references to undercover warfare and making them available through his work at Mie University. *Amagoisan rōjō okite kaki* concerns the siege of the small mountain fortress of Amagoisan. The manuscript is undated, and although Kawakami reckons the attack took place in 1579, this cannot be right because Oda Nobukatsu's invasion entered from Ise not Kōka. The siege must have taken place under the troops of Gamō Hidesato during the second invasion of Iga along the Tamataki route from Kōka in 1581. Kawakami includes the document in his 2016 book *Ninja no okite*.[10]

The work is unique on several counts. First, unlike *Iran ki*, which was composed a hundred years later, *Amagoisan rōjō okite kaki* dates from the time it describes. It is therefore the oldest known reference for the events of the Iga Rebellion and very probably the only contemporary local document about the campaign to have survived the province's destruction. Second, unlike the four raids discussed above, it describes an event within Iga itself. Finally, the 'ninja' element in it is quite different from the four castle raids because shinobi is used as a noun. The 'written regulations' consist of forty-eight clauses, of which three refer to secret operations. The first concerns the out-posting of *shinobi ban* 忍番 (secret guards), whose duty it is to warn the castle of a night attack by the enemy army:

> Item: *shinobi ban*, four men must go out bearing firearms; in the north to the Rairaku Yashiki and the Negoro Yashiki on Kodake, in the east on top of Hachiman, in the south Onosaki, the west Koguchi of Mitaniguchi, they should take up position and conceal themselves in these four locations, and when the enemy come forward for a night attack they must spontaneously discharge two shots to make those within the castle aware. Apart from this they must not fire.

This is an excellent example of surveillance by secret scouts using a simple but effective means of communication back to base to indicate the sector that is being attacked. Beacon fires will then be lit to convey the news back to the main castle, of which Amagoisan is an outpost. The use of the word *shinobi ban* for these brave men seems to be unique to this document, but shortly afterwards the character 忍 appears in a more familiar guise:

> Item, concerning the sending of shinobi 忍び into another province, use and hold to ransom retainers and farmers from the villages and the inhabitants of Hino town in Ōmi, and the villagers and merchants in the environs of Ishibe, but rules are important in this regard. There are allies in the same province, so devise conversations to settle any differences with the Kawai band, the Ōi of Tateoka, the Tsuge band and the Sasaki, and make a fresh start…

This section appears to be the only known use of shinobi as a noun in a pre-conquest account of undercover operations involving Iga. It is interesting because the role given to them is akin to that of the Takeda's *suppa*: extracting information from the inhabitants of an adjoining province, even to the extent of holding them to ransom. Even more interesting is the fact that the adjoining province is Ōmi and more precisely its subdivision of Kōka District. As noted briefly above, Kōka had submitted to Oda Nobunaga before his campaign against Iga began and troops from Kōka are involved in his invasion. Kōka, once Iga's greatest ally, is therefore officially enemy territory, but because of the long association they had enjoyed there are many families there who will retain sympathy and support for the Iga Rebellion and must not be badly treated during the intelligence gathering operation. Instead a meaningful dialogue may be had.

The third use of shinobi in the account appears as, 'When the enemy gather, three horsemen should set out from one of the six units as *shinobi yomawari* 忍夜回り (the secret night watch). In the circumstances it may be desirable to bring along retainers or housemen but if not even one of them is person of good social standing it is useless to take them along'. This suggests that intelligence-gathering requires intelligence. The secret night watch are not casually recruited mountain bandits whose families have to be held hostage, but are more akin to the Hōjō's élite mounted scouts, and the point is also made that if intelligence gathering is carried out it has to be done properly.

All in all, *Amagoisan rōjō okite kaki* is an important addition to our meagre knowledge of authentic secret operations by Iga men prior to the province's destruction. The document may not prove the existence of Iga ninja (as some enthusiasts will no doubt claim), but it shows that even if there was no exclusivity there was definitely local expertise and that its practitioners were called shinobi as early as 1581. They were skilled Iga men who bore the unusual job titles of 'secret guards' and 'the night watch', not unreliable bandits recruited merely to cause mayhem. *Amagoisan rōjō okite kaki* is therefore the oldest document known to exist that links the word shinobi to the area with which it would one day be so firmly associated.[11]

It is an interesting exercise to compare the language used in the above five accounts with the references to Iga warriors found in the later Tokugawa Period war chronicles. For example, there is a reference in *Azai Sandai ki* of 1672 to Iga shinobi no mono from the prominent family of Ban (件の伊賀忍者) taking part in a night raid on Futō Castle in

Northern Ōmi in 1561.[12] Had the work been written a century earlier it would probably have referred to the Iga-shū, with shinobi used only as an adverb, just as it is in the above five accounts that date from the time when the events actually happened. Unlike *Azai Sandai ki*, the earlier five passages are primary sources, and at the very least they appear to support the popular notion that the men of Iga enjoyed a particular expertise in this type of warfare. However, different conclusions may be reached on examining the passages carefully. Shinobi is used in the form of a noun only in the fifth account, so even though the Iga men have expertise, they are not necessarily specialists whose lives are dedicated to shinobi activities as the ninja myth would have us believe. Instead the emphasis throughout is very much upon them operating in a shinobi manner in well-organised military units known as *shū*, a term that provides important clues about Iga's situation prior to 1581.

The *ikki* of Iga and the Iga-shū

At this time in Japanese history the word *shū* usually had the specific meaning of a military unit that was under the command of a *daimyō*, who typically ruled a *kokka* 国家 (domain), 'a political unit defined by the reach of his military and public authority'.[13] All *daimyō* sought to become absolute masters of their territories, aiming to reduce all samurai to vassalage and all farmers to tax-paying workers, so the relationship within a *kokka* was hierarchical and feudal. A *daimyō*'s *kokka* might well have an untidy geographical shape, because it usually consisted of a composite of separate fiefs either held directly by the *daimyō* and his family or indirectly by his closest followers (家来 *kerai*), for whom the European term 'vassal' or 'retainer' is customarily employed. For example, in 1559 the Hōjō of Odawara had serving under them about 500 samurai retainers who were identified using -*shū* as a suffix added to the name of the castle where they would be mustered in times of war such as the Odawara-shū, or indicating a functional military unit like the Go-umamawari-shū (Horse Guards).[14] Does the term Iga-shū therefore indicate a similarly specialised shinobi unit in a *daimyō*'s army?

This is highly unlikely, first of all because contemporary Iga was not part of anyone's *kokka*. The hierarchical *daimyō* model was only one of two possibilities for contemporary governance that had arisen out of the breakdown of central authority following the Ōnin War. The other pattern was for the local samurai to form an *ikki* 一揆 (confederacy or league) when danger threatened. An *ikki* was a voluntary organisation run ideally on egalitarian lines. Some *ikki* had an almost permanent status, of which the prime example was the Ikkō-ikki一向一揆, the massive confederacy drawn from the members of the Buddhist Jōdo Shin sect who defied Oda Nobunaga for ten years from their fortress cathedral of Ishiyama Honganji.

To understand the significance of these alternative models we must backtrack a few decades to examine the normal pattern of government control in the time before the Ōnin War of 1467. The Shogun ruled Japan through his provincially based *shugo* 守護 (governors) or, if the *shugo* chose or was forced to live in Kyoto, through the *shugodai* 守護代 (deputy governors). The provinces consisted then of a patchwork of landholdings held variously by absentee noble proprietors living in Kyoto, religious institutions like Tōdaiji and, most importantly in the context of this book, by small semi-independent landowners sometimes called *kokujin* 国人. The word simply means 'a man of the province' and can be expressed in a more collective sense as *kunishū* 国衆, but its use in contemporary documents can vary enormously. Sometimes it means a man at the highest level of the provincial samurai hierarchy who owned a castle and might even be a *shugodai*. Alternatively, it could mean someone whose jurisdiction was limited to one village, making him little more than a village headman. Another word for *kokujin* was *jizamurai* 地侍 (country samurai), which more clearly identified the *kunishū* as the part-time samurai/farmers that most of them were. One author distinguishes the more powerful castle-owning *jizamurai* from *daimyō* by using a useful analogy from medieval Europe and calling them 'barons', although contemporary Kyoto courtiers and the higher ranks of the samurai class did not regard any *jizamurai* as real samurai.[15] The scorn heaped upon those small landowners from an earlier generation who were casually labelled *akutō* cannot have been very different.

Peace was maintained and war conducted in the provinces via two levels of interaction. The first was between the Shogun and his *shugo*. The second was between the *shugo* and the *kokujin*, and this was the relationship that changed the most following the upheaval caused by the Ōnin War. The resulting Sengoku Period was a time of military opportunity, and in the new age very few absentee *shugo* managed to create their own power bases and become *daimyō* in the provinces they had once supposedly governed. Instead the initiative was taken by locally based *shugodai*, *kokujin* and even erstwhile *ashigaru* (foot soldiers), who saw their chance to seize power in ways ranging from defeating a fading *shugo* in battle, murdering their own commanders or even arranging to be adopted into a well-established family that needed a militarily competent heir. These men had little inclination to recognise the authority of the Shogun except when it suited their own purposes.

In time all the smaller *daimyō* would become absorbed by larger ones, and the first major step towards a genuine reunification of Japan was made by Oda Nobunaga, who combined a lofty contempt for *kokujin* with utter ruthlessness in destroying them. This was particularly noticeable in Echizen Province in 1575, when 12,500 prisoners were taken from the defeated lower-class Ikkō confederacy. Most of them were executed in cold blood, and in a letter to his representative in Kyoto Nobunaga wrote:

> Within the town of Fuchū dead bodies lie everywhere with no empty space between them. I would like you to see it. Today, hunting mountain by mountain, valley by valley, I have to complete the task of seeking out and exterminating them.[16]

If opponents submitted to Nobunaga they might be better treated, and this treatment was noticeably more generous the higher his victims were within the warrior hierarchy. In 1582, for example, Nobunaga issued regulations for the newly conquered provinces of Kai and Shinano. Item Three stated, 'The loyal shall be left in place. As for the rest, kill or banish any samurai causing mischief', although item 5 read, 'Treat the provincial samurai with courtesy. For all that never be remiss in your vigilance'.[17]

Iga Province's local confederacy was called the Iga sōkoku-ikki 伊賀惣国一揆 ('the whole Iga provincial league'), and there are records of them fighting in a military unit called the Iga-shū as early as 1482. In 1485 the Iga *kokujin* are to be found defending a castle in Yamato Province on behalf of the Hatakeyama family and in 1492 they are involved with the Rokkaku family of Ōmi Province.[18] By the time of the Kasagi attack in 1541 the Iga Confederacy were riding high, because even though the rule of them by the Iga *shugo*, a hereditary post kept within the Nikki family, had become increasingly irrelevant, their absentee aristocratic ruler had not been replaced by a rising *daimyō* family like the Hōjō. Instead the province was still divided among strong-minded *jizamurai* who had not passed under anyone's vassalage. These were the men who fought in the Iga-shū as the Iga Confederacy, and as such they defied Oda Nobunaga and his son over the course of the two campaigns that ultimately led to Iga's downfall in 1581.

The mutual need for survival in the face of threats from *daimyō* and their rapacious armies was usually sufficient to bind the members of a provincial *ikki* together. In some provinces an extra factor, such as affiliation to a particular sect of Buddhism, strengthened the bonds. Pierre Souyri, who accepts the Iga ninja tradition uncritically, believes that the supposed ninja tradition provided the 'extra glue' in the local *ikki* that held the Iga Confederacy together.[19] There is, however, no evidence for this, nor is there any hint of it in the Iga Confederacy's written constitution. This was a document typical of the better-organised *ikki*, to which the *kokujin* leaders subscribed by making solemn pledges, and it is fortunate in view of the destruction of Iga in 1581 that its details have survived. In the document military service is required from any man aged between fifteen and seventy. There is also a commitment for the *ikki* to respond as one if the province was invaded, and a specific intention to cooperate with their friendly neighbours across the border in Kōka.[20]

Souyri also errs when he regards Iga as politically unique, because a comparison with other areas demonstrates that Iga was no different from any province where *daimyō* rule was not the dominant force and the combined strength of the *kunishū* allowed its barons to preserve the province for themselves in spite of inroads made by more powerful neighbours. Southern Japan provides the two examples of Satsuma and Higo Provinces (the modern prefectures of Kagoshima and Kumamoto respectively). Higo Province was for decades a battlefield for the three great powers of Kyushu: the Shimazu, the Ōtomo and the Ryūzōji, whom the local barons fought for and against in a confusing series of voluntary and forced alliances. Like Iga Province, Higo and Satsuma provided mountains as places of refuge. The phrase 'a domain so completely surrounded by a broad river and complicated mountain systems must have been singularly impregnable against invasion', refers not to the fanciful comparison I once made between Iga and Switzerland,[21] but to the situation of the *kunishū* barons who held sway in Satsuma.[22]

The big difference between Iga and Satsuma, of course, involved their comparative locations. Kagoshima is 900km from Kyoto even via modern expressways. The Iga-shū confederates were based near enough to Kyoto to influence its politics indirectly by lending their support to factions with some direct influence, and they always had their mountains to provide a refuge if retreat became necessary.

Nevertheless, the *Tamon'In nikki* account appears to undermine any notion of independent action when it states that the Iga-shū attacked Kasagi as a result of an order from the Shogun to their provincial *shugo*. That, and the fact that they were fighting outside their own province in 1541 has been interpreted in the past as evidence for mercenary activity on behalf of the Tsutsui at the *shugo*'s command, so the Kasagi operation is cited as proof that the Iga shinobi no mono sold their specialised services to outside lords. Yet that is neither likely, nor in any way necessary, to explain the Kasagi situation. Kasagi Castle was quite close to Iga. By modern roads it is a mere 20km from present-day Iga City to Mount Kasagi. Adding to this the fact that in 1541 boundaries were often defined not by province, but by what territory could be defended, such an involvement in someone's dispute by his neighbours is by no means uncommon, particularly when the warring neighbours are rival *daimyō*. Just like the barons of Higo the Iga *ikki* would fight for or against a particular *daimyō* when it was in their interests to do so. The Tsutsui of Yamato, who are mentioned in the first three accounts above, were also no strangers to the Iga *kokujin*. In 1581 Tsutsui Junkei would play a role in the invasion of Iga and Junkei's son Tadatsugu would become Iga's *daimyō* in 1585. There is therefore no need to assume that in 1541 the Iga-shū were acting as politically disinterested mercenaries seeking only financial gain from a willing employer. It was very much in their interests to be involved with a powerful neighbour. To succour friends and defeat enemies is fully sufficient to describe the Iga-shū's operations over the border in Yamato, Ōmi and Yamashiro. In 1541 their *shugo* would have been pushing at an open door.

The fourth account quoted above is somewhat different because the operation at Sakaibe was clearly designed to defend Iga by attacking Oda Nobunaga's allies outside the province. The priest's admiration for their shinobi expertise is crystal clear, but even though the five accounts are the only ones covering secret operations, several more have survived in which the Iga-shū are found to be fighting more conventionally, but are no less admired. *Shinchō-Kō ki*, for example, records that in the battle of Yasugawa in 1570 Nobunaga's men killed '780 of the finest samurai from the Iga and Kōka-shū'.[23] In 1573, 'skilled archers from Iga and Kōka' are found assisting the Ikkō-ikki of the Nagashima delta against Nobunaga, whose army was in retreat. They 'let loose a hail of arrows that felled countless men'.[24] The activities of the Iga-shū were therefore not confined to secret attacks and have to be seen in a wider context of an overall high-level expertise at warfare that

48

attracted attention from friend and foe alike. This was a very necessary accomplishment if Iga and its *ikki* were to survive at all in the midst of an extremely hostile environment where attack was often the best means of defence.

The picture that emerges for Iga's military operations prior to 1581 therefore provides an alternative to the models of super-samurai or thieves for understanding the word shinobi. The expression may appear in the authentic accounts primarily as an adverb, but when the character *nin* exists in a noun compound such as *shinobi ban*, the operations are clearly being carried out not by casually recruited criminals, but as part of a well-organised *jizamurai* confederacy. Its members are still comparatively lowly in the social hierarchy, but their operations are not mercenary activity, nor are sneak attacks the be-all and end-all of Iga's military expertise. Shinobi activity is just one of a wide range of military skills possessed by these rugged fellows, a genuine expertise that is closely related to their social and political situation. In this model the origins of the ninja really are to be found in the independent samurai/farmer communities so beloved of the movie industry.

It must, however, be reiterated that Iga and Kōka were far from being unique either in their undercover operations or their social situations. Higo Province, which was mentioned above for its political and geographical similarity, had several staunch communities of *jizamurai* who also displayed great skills in warfare. Higo's independence came to an end in 1587 at the hands of Toyotomi Hideyoshi. During the final act of rebellion by the Higo *kunishū* we come across several shinobi incidents which, if they had been performed by men from Iga, would have been added to their tally of 'ninja' operations. Supply columns were harassed and castles were attacked,[25] culminating in a siege of the rebel stronghold of Tanaka Castle.[26] As the operation against the tiny fortress wore on, Hideyoshi's frustrated commander used an arrow to convey a message to Hebaru Chikayuki, the one member of the castle command whom he believed to be susceptible to negotiation. The ploy succeeded, and Chikayuki persuaded a retainer called Usono Kurandō to kill the castle commander Wani Chikazane.[27] *Wani Gundan*, which chronicles the events, reads:

> Kurandō agreed and in the middle of the night of the 6th day of the 12th month he entered secretly (*shinobi no itte*) into Chikazane's private quarters and killed him, cutting off his head he escaped into the enemy lines. Chikayuki sent a signal by means of a beacon and

the attack was launched. At the same time Hebaru Chikayuki set fire to the inner bailey and because of a strong wind the castle was engulfed in flames.[28]

Needless to say, the assassin at Tanaka was neither a ninja nor from Iga. Kurandō was instead a disaffected subordinate in the garrison hierarchy who had been passed over for promotion and killed his commander secretly in an act of revenge. He was promised a reward for the deed, but his treachery found him no friends among the victorious besiegers and he died a beggar.[29]

More secret activity took place in Higo two years later when the unrest spread to the Amakusa Islands. In October 1589 Konishi Yukinaga despatched 3,000 troops against the rebel baron Shiki Rinsen.[30] The army advanced without hindrance as far as the island of Shimojima, unaware that they were being led into a trap. There Rinsen caught them in a surprise attack and massacred them.[31] Yukinaga then led a much larger army against them, so Rinsen sent a secret envoy to Yukinaga's ally Ōyano Tanemoto to discuss peace terms. The man is referred to as an (unglossed) 忍者, and although it is tempting to read the compound as ninja and assume he is a secret agent, it obviously indicates no more than that he was on a covert mission because his intention was to discuss peace.[32]

The members of the Iga Confederacy, who resembled the *kokujin* of Kyushu in everything except their physical location, were involved in similar activities over a period of many centuries, although there do not appear to be any records of peace missions being undertaken by their members. Instead they seem to have been engaged in a delicate balancing act both for influence and survival in their tiny strategic enclave, a task for which undercover warfare was a necessary accomplishment. In this they were supported for many years by their neighbours from Kōka, with whom they would forever be linked and to whose history we will now turn.

Chapter 4

Kōka and Iga

The military pressures exerted upon Ōmi Province in general and the district of Kōka in particular were very different from those inflicted upon Iga, even though the final result would be the same: the collapse of local independence and the imposition of *daimyō* control. Kōka District lay in a highly strategic position and had always provided a bridge between Iga's greater isolation and key lines of communication running through Ōmi. Kōka was therefore far from being either inhospitable or remote, as some ninja enthusiasts have claimed. On the contrary, Japan's famous road, the Tōkaidō, ran straight through it, taking in three of the places that would be immortalised during the Tokugawa Period by the print-maker Hiroshige as the 'Fifty-Three Post Stations of the Tōkaidō'. Station 51 was Ishibe in modern Kōnan City. To the east lay Station 50 at Minakuchi in Kōka. After leaving Minakuchi the eastbound Tōkaidō passed through Station 49 at Tsuchiyama before heading up to the Suzuka Pass, which took the road into Ise through the formidable Suzuka Mountains, the highest peak of which is 1,247 metres above sea level.

The historical Kōka area consists of two main valleys and several smaller ones traversed by rivers flowing down from the mountains that separate it from Iga. At one point, Mount Aburahidake, the provincial border passes through the mountain's summit. The two main rivers of Kōka are the Yasugawa and the Somagawa, which converge in the area of modern Minakuchi City. The sites of about a hundred medieval fortifications have been discovered dotted around the hills above them, evidence of small-scale yet extensive military activity.[1] A typical arrangement is found at Takigawa Castle, which was associated with the Ōhara family. There are three individual fortified places. The *kyōjō*

51

(main castle) lies across the valley from the Tendai temple of Rakuyaji. Adjacent to it is the *saijō* (western castle), while across the valley is another *bunjō* (subsidiary castle) in a pattern of plural castles that is typical of the area. Each is built on the *yamashiro* (mountain castle) model, and the only distinguishing feature of the main castle is that it is larger in area than the other two.[2] The high-ranking *jizamurai* who owned castles like these were the 'barons' of Kōka, and at least one of them held state within his own small valley as if he were a *daimyō*. He was Wada Koremasa (1530–1571), whose seven fortified places included Wada Castle and no fewer than five *bunjō* on adjacent hills, each built within sight of at least two of the others. The largest fortified site of all lies on a separate hill and is identified as the site of the *kubō yashiki*, the Shogun's mansion. It was erected or at the very least extended for a distinguished fugitive, as we will see below.[3]

The existence of numerous small castle clusters like these implies that Kōka was organised on a similar *ikki* pattern to Iga, and this is indeed the case, although the terminology used was different. The *jizamurai* who garrisoned the Kōka castles appear in the early Sengoku records in family groups called *dōmyōchū* 同名中 or *ichizoku shūdan* 族集団. Both words indicate that they possessed the same surname, so 'clan' is probably the best translation. The smaller-scale landholders made up the Kōka *samurai-shū* 侍衆, a word which appears in a document of 1475 that bears the names of the six clans of Tomita, Masuda, Shiotsu, Nishioka, Kitano and Nakagami. It refers to donations they made to the shrine of Aburahi Daimyōjin, the tutelary deity of Kōka.[4] Just as in Iga, their social structure meant that the Kōka barons were involved in their own version of an *ikki* that some writers have dubbed 'Kōka republicanism' (甲賀共和制 *Kōka kyōwasei*).[5] Ōmi province was too large for a whole-province sōkoku-ikki to be feasible, so the historical expression for the Kōka Confederacy was the Kōka-gun Chūsō 甲賀郡中惣, an assembly of the *dōmyōchū* clans of Kōka. The name may be found in historical records in 1571 that relate to mediation in a dispute between the Shingu and Yagawa Shrines and the Handōji, Kōka's main Shugendō temple. In 1584 the names of Ōhara, Hattori, Mochizuki, Ikeda, Ukai, Ichiyaku, Taki, Saji, Takamine, Ueno and Oki are noted as members of the Kōka Confederacy.[6]

As well as being too large for a whole-province league, Ōmi was also too diverse and, most important of all, too strategically valuable for Kōka-style 'republicanism' to apply throughout the province. Ōmi had Lake Biwa as its centre, which meant that all major communications were crammed within the flatlands between the lake and the encircling mountains. These communication routes were vital to Japan. At Kusatsu the westbound Tōkaidō that had come through Kōka was joined by the Nakasendō that linked east and west through the central mountains rather than the sea coast. Other major roads diverged from them both, making Ōmi a pinch point for Japanese trade and traffic, and no fewer than 1,300 fortified sites have been identified within its provincial borders. The challenge posed by Ōmi's strategic situation would be realised time and again in Japanese history, because anyone who controlled Ōmi Province could control the life of Kyoto, whose citizens could not survive more than a few weeks if food supplies were cut off. At the same time, of course, any ambitious *daimyō* who sought to advance on Kyoto from the east could be stopped by determined warriors in Ōmi.

Oda Nobunaga and Kōka

Among the determined warriors of Ōmi were the Kōka Confederacy, who were assisted by their cross-border allies the Iga Confederacy. Both were to suffer at the hands of the unifier Oda Nobunaga, although the means whereby the power of the Kōka *ikki* was curtailed would be very different from Iga's own bitter experience. In part this was due to Kōka's situation, because Ōmi Province was large enough to house not only *jizamurai*, but also two *daimyō* who were inferior to Nobunaga only in their ambitions. In southern Ōmi were the ancient Rokkaku family. They were descended from the Sasaki, who had supplied several heroes during the Gempei War including Sasaki Takatsuna, who famously swam his horse across a river at the battle of Uji in 1184.[7] Sasaki Takatsuna's great-nephew Yakutsuna took the name of Rokkaku during the thirteenth century.[8] By the time of Nobunaga the Rokkaku had made a successful transition from *shugo* to *daimyō* and now ruled their territories in the person of Rokkaku Yoshikata (1521–1598). In 1562 Yoshikata shaved his head and took the Buddhist name of Jōtei. He then officially ceded his domains to his son Yoshisuke (1545–1612), but in view of how often Rokkaku Jōtei's name is mentioned he continued to exercise considerable influence within the family.

Rokkaku Jōtei ruled his domain from Kannonji Castle, which was built on a distinctive mountain that soared out of the flat plain near Lake Biwa. He also had an important base just inside Kōka at Ishibe, where a castle commanded by the Aoki family guarded the Tōkaidō road and the Rokkaku family temple of Chōjūji protected their ancestral spirits.

53

The local Kōka *ikki* enjoyed a good relationship with the Rokkaku and would willingly fight for them. The earliest recorded example of the alliance dates from 1487. Rokkaku Takayori, the *daimyō* at the time, had been so aggressive against his neighbours that the Shogun Ashikaga Yoshihisa launched a campaign against him called the Chōkyō no Ran from the year period in which it happened. Takayori received help from Iga and Kōka in return for recognising their land ownership. The Shogun's invading forces set up a base at Magari (in modern Rittō City, Shiga Prefecture), where they were assaulted by unexpected night attacks from the Iga/Kōka men that drove back the Shogun and hastened his premature death. This was the conflict that *Ōmi Onkoroku* 淡海温故録 described as the first operation by 'the Iga-Kōka shinobi no shū' who 'proved themselves in front of huge armies assembled from all over Japan'. The Kōka men certainly used surprise hit-and-run tactics to disrupt the Shogun's invasion, so the only embellishment brought in by *Ōmi Onkoroku* is probably the shift in the use of shinobi from an adverb to a noun.[9] That has been enough to ensure that the Magari Incident has been cited in all subsequent popular books as Japan's first ninja operation. What is known for certain is that twenty-one families from Kōka were officially honoured by the Rokkaku for their unconventional part in the conflict.

The Rokkaku family's rule only covered Southern Ōmi, and by the sixteenth century much of Northern Ōmi was ruled by the Rokkaku's unsympathetic rivals, the Azai. Their territories met at Hida Castle in Aiwa-gun, a place that would be contested fiercely on at least two occasions, and the rivalry between the families went back two generations.[10] In a way typical of the Sengoku Period Azai Sukemasa (1495–1546) had revolted against his overlord Kyōgoku Takaie in 1516 and set himself up in the castle of Odani. As the Kyōgoku were also descended from the Sasaki family Rokkaku Sadayori (Jōtei's father) took the incursion as a personal insult and led an unsuccessful attack on Odani in 1518. After that the rivalry passed down the generations with Rokkaku Jōtei fighting Azai Hisamasa (1524–1573) until Hisamasa's own retainers persuaded him to retire in favour of his son Nagamasa (1545–1573). Young Nagamasa then took the fight to the Rokkaku, and it is at around this time that we begin to read of involvement from Iga and Kōka against the Azai on the Rokkaku's behalf. During the early 1560s Rokkaku Jōtei took on a more formidable enemy than the Azai when he came to blows with Oda Nobunaga. It proved to be a long and deadly war, which began with a succession dispute for the post of

Shogun, a serious development for which a short preliminary explanation is needed.

The history of the Ashikaga Shoguns is peppered with references to them fleeing, being replaced, being exiled or even being assassinated. By 1545, when the 13th Ashikaga Shogun Yoshiteru (1535–1565) was appointed at the age of ten, the pattern had become so familiar that he may not have been too surprised when he was forced to flee from Kyoto. He returned eight years later under the thumb of the malignant Miyoshi Chōkei (1523–1564) and his supporter Matsunaga Hisahide (1510–1577). To try and get rid of them Yoshiteru sought help from outside, and one of his ploys was to appoint the up-and-coming young *daimyō* called Oda Nobunaga to the post of *shugo* of Owari Province in 1559. The title meant little in military terms to Nobunaga. Its value lay instead in giving Nobunaga official approval for the complete takeover of his home province that he would accomplish during that year.[11]

No one, not even Oda Nobunaga, could have anticipated the surprise night attack against the Shogun's palace that Miyoshi and Matsunaga launched in 1565. As an accomplished swordsman who had been a pupil of the legendary Tsukahara Bokuden, Shogun Ashikaga Yoshiteru took on his assailants, but when he was mortally wounded he crawled into an adjacent room and committed *seppuku*. After this successful part of the coup the plotters' plans started to fall apart. A lack of imperial support and dissension among the conspirators meant that their nominee for the position of fourteenth Ashikaga Shogun, Yoshiteru's distant cousin the infant Yoshihide, was refused investiture. Frustrated in this respect, they lighted instead upon Yoshiteru's younger brother Ashikaga Yoshiaki, who was living at the time as a monk in Nara. The plotters sought out Yoshiaki and pledged that no harm would come to him, but their assurances were not believed and Yoshiaki slipped away from Nara before they could murder him.

The fugitive Shogun-in-waiting fled eastwards into Kōka, where he found refuge with the baron Wada Koremasa.[12] Wada Castle may have been able to withstand an attack by Miyoshi and Matsunaga, but Yoshiaki had not run away only to avoid them. He wanted to become Shogun in his own right and Wada Koremasa was not the sort of strong provincial *daimyō* who could deliver on that promise, so Yoshiaki moved on to Yashima in Ōmi and the protection of Rokkaku Jōtei, whom he believed would escort him to Kyoto to claim his throne.[13] Jōtei proved unwilling to risk the good relationship he enjoyed with the Miyoshi faction and soon sent Yoshiaki on his way again. As *Shinchō-Kō ki* puts it so well, 'The tree Yoshiaki had sought for shelter was letting the rain through'.[14]

The increasingly desperate Yoshiaki then sought help from other *daimyō*. Asakura Yoshikage (1533–1573) of Echizen Province gave him refuge, but rejected any notion of helping Yoshiaki become Shogun. Oda Nobunaga, however, indicated his support because he realised that Yoshiaki could help serve his own ambitions. The wretched Yoshiaki probably knew that too, but he was eventually driven into Nobunaga's arms, and among the envoys sent by Nobunaga to escort Yoshiaki from Echizen in 1568 was Kōka's Wada Koremasa. Koremasa's continued support for Yoshiaki had inevitably pushed him into the pro-Nobunaga camp, and later that same year he was given the castle of Akutagawa in Settsu Province as a reward, so the Wada's association with Kōka came to an end.[15]

By the summer of 1568 Nobunaga was ready to march on Kyoto and install Ashikaga Yoshiaki as Shogun, but he had a major problem because Southern Ōmi was enemy territory. Nobunaga's army expected to pass peacefully through Northern Ōmi because it was ruled by Azai Nagamasa, who had helpfully become Nobunaga's brother-in-law. In hostile Southern Ōmi Nobunaga would have to face the distinctly unsympathetic Rokkaku Jōtei, and even a week-long set of negotiations, during which Jōtei was promised the position of Governor of Kyoto, failed to persuade him to let Nobunaga's army through. Not only did Jōtei refuse to provide a safe passage through Ōmi, an act that would have normally been guaranteed by the provision of hostages, he declared his support for the rival Ashikaga Yoshihide, the child nominee of the Miyoshi faction who had finally been proclaimed Shogun. With no safe conduct promised, Oda Nobunaga responded by invading Southern Ōmi with an army of 60,000 men.[16] Kannonji Castle soon fell and over the next month Nobunaga's armies crushed all opposition.[17] The young Shogun Yoshihide was killed during the fighting and his sponsors Miyoshi and Matsunaga were forced to pay homage to Nobunaga. On 7 November 1568 Nobunaga's puppet Ashikaga Yoshiaki was installed as the fifteenth, and as it turned out last, Ashikaga Shogun.[18]

Following their shock defeat, Rokkaku Jōtei and his son fled first to Kōka and then took refuge on the holy mountain of Kōyasan.[19] They had been unable to prevent Nobunaga's march on Kyoto, but were now determined to make life as uncomfortable for him as possible. From Kōyasan and from successive bolt-holes they continued to organise resistance against Nobunaga in a largely guerrilla war. The Kōka and Iga confederacies responded readily to the call to arms from the dispossessed Rokkaku. Their intimate knowledge of the area and renowned fighting skills meant that Southern Ōmi was never entirely safe for Nobunaga, as was illustrated best of all in 1570. Nobunaga was campaigning against the Asakura of Echizen Province when his brother-in-law Azai Nagamasa turned traitor and attacked him in the rear, forcing Nobunaga's army into a fighting retreat. Southern Ōmi provided the direct route back to Kyoto, but instead Nobunaga had to make his way round the north-western shore of Lake Biwa guided by local sympathisers, thus avoiding any confrontation with Rokkaku Jōtei and his troops from Iga and Kōka.

As to the genuineness of the threat from that quarter, *Shinchō-Kō ki* says that, 'An armed confederation had already risen against Nobunaga, put the village of Heso to the torch, advanced on Moriyama, and set fire to its southern approach'.[20] The location of Moriyama beside Lake Biwa indicates that on this occasion the *jizamurai* of Iga and Kōka were assisting the Rokkaku by raiding across Nobunaga's main lines of communication, but disaster was soon to come upon the confederates. On 6 July 1570, as they moved down the Yasugawa under the leadership of Rokkaku Jōtei and his sons, they were intercepted by Nobunaga's army under Shibata Katsuie and Sakuma Morimasa at the village of Ochikubo. This was the battle of Yasugawa noted earlier, at which '780 of the finest samurai from the Iga and Kōka-shū' were killed, along with 'the Mikumo father and son Takanose and Mizuhara'. This quotation from *Shinchō-Kō ki* is the only written source for what must have been a major fight.[21] A casualty list of 780 was enormous considering that the Iga/Kōka army cannot have been very large. It was a savage blow to the fighting strength of the two *ikki*, and in fact we hear nothing of either Iga or Kōka in *Shinchō-Kō ki* for another three years.[22] There may also have been some defections to Nobunaga from within their ranks, because later in 1570 Rokkaku Jōtei and his son are found attacking Bodaiji, a fort just inside Kōka held by the Mikumo family. The survivors from the Mikumo had clearly passed over to Nobunaga's side. Bodaiji Castle, which cannot be precisely identified, was not far from the Rokkaku outpost of Ishibe and should have fallen easily, 'but the Sasaki were short of men and not ready for battle'.[23] It was the lowest point for the *ikki* in their war against Nobunaga.

At about the same time a very different attack on Oda Nobunaga took place when a remarkable attempt was made on his life by a monk called

Sugitani Zenjūbō, who was obviously a crack shot with a musket. He lay in wait for Nobunaga and fired two bullets at him (presumably by having two separate guns ready primed and loaded), but both bullets merely grazed their noble victim.[24] According to *Shinchō-Kō ki* the man was 'hired' or 'engaged' by Rokkaku Jōtei, which has been taken as meaning that he was a mercenary shinobi assassin from Iga. Whatever Zenjūbō was, he met a horrible death three years later. Zenjūbō was apprehended by Nobunaga's men and interrogated about how he had managed to launch his ambush. He was then executed 'in a way Nobunaga had specially designed. He was buried upright to his shoulders, and then his head was sawn off. Everyone, high and low, was very satisfied with this punishment'.[25]

Rokkaku Jōtei continued the fight against Oda Nobunaga in Southern Ōmi, but the massacre at the battle of Yasugawa had meant that the Kōka and Iga confederacies were temporarily unable to assist him. In Northern Ōmi, however, the Rokkaku received help from a different source when Ikkō-ikki members based in Nagashima tried to cut Nobunaga's lines of communication. 'They were farmers and therefore of no account', says *Shinchō-Kō ki* dismissively, so Nobunaga burned their villages and killed all the inhabitants. The Ikkō-ikki were undaunted and at one battle brought about the death of Nobunaga's younger brother. Prompted perhaps by this serious development, Nobunaga concluded a very welcome truce with his lesser enemy Rokkaku Jōtei on 19 December 1570, thus allowing him to concentrate on his long war with the Buddhist sectarians.[26]

Nobunaga's truce with the Rokkaku lasted until 1573, by which time the political situation had changed considerably. Over the course of the three years a split developed between the independent-minded Shogun Ashikaga Yoshiaki and Oda Nobunaga, the man to whom he owed his position. Seeking to be rid of the yoke, Yoshiaki had sought alliances with Nobunaga's enemies and in 1573 the animosity turned into armed hostility. The Rokkaku gleefully took the Shogun's part, along with their allies from Iga and Kōka, who had rebuilt their strength in the three years since Yasugawa. As the Shogun's loyal followers they were now firmly aligned against Nobunaga at the very highest level of state politics. Mindful of the importance of communications between Kyoto and Ōmi, Ashikaga Yoshiaki established a castle at Ishiyama beside the strategic 'neck' of Lake Biwa. Men from Iga and Kōka formed part of the Ishiyama garrison, although their defence was feeble. Records relate that the fort was only half finished when Nobunaga attacked it and the defenders ran away, pleading for mercy. The place was immediately demolished.[27] Later in that same year of 1573, 'skilled archers from Iga and Kōka' are also found fighting against Nobunaga in Northern Ōmi Province, where they are assisting the Ikkō-ikki of Nagashima. They 'let loose a hail of arrows that felled countless men. Because of the heavy rain, the firearms of either side were useless'.[28]

After more skirmishing with Nobunaga and the burning of part of Kyoto in retaliation Ashikaga Yoshiaki took refuge in a castle called Makinoshima near Uji in Yamashiro Province.[29] He left a garrison behind in Nijō Castle, his stronghold within the city of Kyoto. Nobunaga wanted to respond, but knew that his advance might once again be hindered by the Rokkaku. To outflank them Nobunaga decided to attack Kyoto by sailing along Lake Biwa. He set up a base at Sawayama (modern Hikone City) and ordered the construction of three large ships that could transport his troops across the lake instead of marching through Ōmi. In late summer 1573 Nobunaga landed at Sakamoto and advanced quickly against Nijō Castle, which soon surrendered. Having consolidated his position in the capital Nobunaga trod the historic road from Kyoto to Uji, forced a crossing of the Ujigawa and attacked Makinoshima. There Ashikaga Yoshiaki made his last stand against Oda Nobunaga, but after only one day of fighting he prudently surrendered and the two and a half centuries-long Ashikaga Shogunate came to an ignominious end.[30]

To round off his military triumphs of 1573 Nobunaga completed the conquest of the Asakura and Azai families, and at a banquet held for his Horse Guards at the New Year celebrations of 1574 the *piece de resistance* of the celebration was the display of the heads of their three leading members, tastefully lacquered in gold. In that same year of 1574 he finally overcame the Rokkaku and their Kōka allies. The independent Kōka Confederacy thus disappeared forever, and if a document preserved by the Yamanaka family is to be believed, the event can be dated precisely as the fifth day of the third lunar month (27 March 1574), when the 'remnants' of the Kōka *jizamurai* surrendered to Nobunaga.[31] The Rokkaku family were not far behind them. Jōtei and his son had taken refuge in their castle of Ishibe in Kōka. Nobunaga gave orders for it to be besieged, but on 3 May 'under cover of a rainy night, Sasaki Jōtei abandoned Ishibe Castle'.[32] He submitted meekly to his new overlord, and from that time onwards the Rokkaku father and son lived in Nobunaga's court and the whole of Ōmi Province became Oda territory.[33]

Nobunaga then began a programme of destroying the numerous small castles in Kōka, but apart from that (a practice commonly applied to all defeated territories) Nobunaga behaved towards Kōka with relative generosity.[34] Some families fled Kōka for service elsewhere, but there were no large-scale massacres and no attempt was made to bring someone in from outside to rule them.[35] Their new *daimyō*, after all, was Oda Nobunaga himself, so the expression 'Kōka-shū' lived on as a military unit within Nobunaga's army, and as long as they stayed his loyal vassals and caused no trouble that was how they would continue. It is therefore as the Kōka-shū that they appear within the *Shinchō-Kō ki* army list for the second invasion of Iga Province in 1581.[36] Kōka, to whom the Iga *ikki*'s constitution had once pledged unquestioned cooperation, was now fighting for Oda Nobunaga rather than against him. The great alliance that had sustained them for over a century had been broken and turned into confrontation.

In 1576 Nobunaga stamped his authority on Ōmi Province in a very dramatic fashion by choosing a place called Azuchi as the site for his new headquarters. The location of Azuchi Castle on the shore of Lake Biwa allowed Nobunaga to control Kyoto along with Ōmi. The site was not far from the ruins of the Rokkaku's sprawling *yamashiro* of Kannonji. This may have been an acknowledgement of the Rokkaku's strategic vision, but it was there that any notional homage ended. The palatial Azuchi Castle was no earth-and-timber mountain castle like Kannonji, but a new style of building that set the tone for the great Japanese castles that were to come after it and to which their builders would aspire. Azuchi's strength and overall appearance was to stun the eyes of European visitors who were familiar with the Renaissance defences of Rome. No expense was spared on the huge edifice with its innovative seven-storey keep, and neither was sentiment. Regardless of the feelings of the Rokkaku family, Nobunaga added insult to injury by dismantling most of the family's memorial temple of Chōjūji at Ishibe, a place that had been the pride and joy of Rokkaku Jōtei. It was not destroyed. Instead Nobunaga reassembled the bits of Chōjūji at Azuchi to create the temple of Sōkenji.[37] Most of Azuchi Castle was burned down when Nobunaga died in 1582, but the three-storey pagoda and the gate of Sōkenji survive to this day.

The buildings that made up the Sōkenji would have been sadly familiar to any of the Kōka-shū who lived in Ishibe, and on New Year's Day 1582 their commanders had a chance to see it again. The former *ikki* leaders were now so closely associated with Nobunaga that in

Shinchō-Kō ki they are mentioned in the same sentence as his élite Horse Guards when 'the major and minor lords from the neighbouring provinces as well as the fraternal branches and their retinues' came to visit Nobunaga in his fortress palace. Amusingly, Nobunaga effectively charged for admission, because he 'told major and minor lords alike to bring along one hundred copper coins each as a courtesy fee... After admiring the stage constructed at the Bishamon Hall of the Sōkenji, his guests passed beyond the Front Gate'. At the conclusion of the visit, 'He had them stand at the entrance to the stables, where he graciously collected the ten *hiki* courtesy fee from each directly with his own hands before throwing the coin behind him'.[38] For this small donation the former Kōka *ikki* were able to see a temple they would have remembered and recall a world of independence that they had now lost forever.

Chapter 5

The Pacification of Iga

The process whereby the *jizamurai* of Kōka lost their independence was one that was to be repeated throughout Japan over the next two decades. Iga would follow Kōka within seven years, although Iga's experience was to be very different and very painful. As an earlier chapter made clear, the members of the Iga Confederacy enjoyed considerable expertise in warfare and often defended their province by fighting outside its borders, but these pre-emptive defensive measures came to an end with the first and second Tenshō Iga no ran 天正伊賀の 乱 ,, the 'Iga Rebellions of the Tenshō Era'. The title is somewhat misleading, because Iga was defined as rebellious only by Oda Nobunaga, the man whose army invaded it twice on the pretext of quelling a revolt. The first incursion was carried out by Oda Nobunaga's son Oda Nobukatsu in 1579 and was a humiliating failure. The second was directed by Nobunaga himself in 1581 and crushed the Iga warriors by the application of overwhelming force.

There are five primary sources for the Tenshō Iga no Ran. The first is a very brief mention that occurs in *Tamon'In nikki*, and thanks to Kawakami we now have the short but vitally important *Amagoisan rōjō okite kaki*, with its revealing account of shinobi in action during the second invasion. A longer account makes up a section of Ōta Gyūichi's historically reliable *Shinchō-Kō ki*. As with the rest of the work the Iga Rebellion account glorifies Nobunaga's achievements, but is otherwise dispassionate. It begins with Oda Nobukatsu's failed campaign in Iga and his father's furious reaction. The account of the 1581 invasion describes several battles, listing the taking of heads and the demolition of castles. It concludes with Nobunaga's inspection of the conquered province and the names of its new overlords. Nobunaga's army then moves on to a new target.[1]

A hundred years later, in 1679, the Iga scholar Kikuoka Jōgen (1625–1703) wrote *Iran ki* 伊乱記 (Chronicle of the I[ga] Rebellion). It is most readily accessible in a version called *Kōsei Iran ki* that was edited by Momochi Orinosuke and printed in Iga-Ueno in 1897.[2] His editing involved the addition of *furigana* notes and a modification of the seventeenth-century written style. In 2006 a group of scholars in Iga City published *Iran ki* in its original form based on the oldest manuscripts.[3] Bound in with this edition of *Iran ki* is Kikuoka's *Iga Kyūkō* 伊賀旧考, which covers the same time period but in a less romanticised and more matter-of-fact style. That it was written after *Iran ki* is shown by the inclusion of a list of names noted as being taken from *Iran ki*.[4] The manuscript of *Iga Kyūkō* survives in two versions, one of which bears the date of Bunroku 12 (1699).

All five sources will be used to construct the following narrative. *Iran ki* has to be treated with some caution because it belongs among the category of exaggerated Tokugawa Period works and is written in the lively style of a medieval *gunkimono*, with a considerable degree of romantic and heroic exaggeration. This factor has of course been responsible for the vital role *Iran ki* has played in providing the building blocks for the ninja myth and supplying the plots for several important ninja films along the way. No story has ever done more for the Iga ninja than *Iran ki*, but even if the possibly inflated accounts of individual heroics are stripped out, its story of the province's opposition to the all-conquering Oda Nobunaga is one of the greatest tales of resistance to an invader in the whole of Japanese history.

To create *Iran ki*, Kikuoka probably drew upon a genuine wealth of oral tradition that survived the wars and which we now know is supported by the independent documents discussed earlier. His descriptions of battles are very exciting, and running through *Iran ki* is a tangible pride in Kikuoka's home province and its great martial tradition. Few places in Japan can claim that their local samurai ever inflicted a defeat upon Oda Nobunaga's army, and to Kikuoka the Iga shinobi 伊賀忍 of his day who were serving in Edo Castle were the direct successors of the heroes of the Iga Rebellion.[5]

Kikuoka's pride becomes evident very early in the narrative when he makes a classic statement that was to be echoed over the course of many centuries: 'They have a supernatural power as spies, so no matter how secure a fortress is, they can enter it secretly, and in other provinces too the Iga shinobi have been found to be useful'.[6] *Iran ki* therefore has the paradoxical role of expressing pride in a supposed local tradition, while simultaneously shaping it, although the nomenclature we now associate with ninja is at a very early stage in its development. The word I translated as 'spies' in the above quotation appears as *kanchō* in the original, but is glossed by Momochi to be read as shinobi,[7] while the expression 'Iga shinobi' (written as 伊賀忍び later in the passage) means the men of Kikuoka's own time in Edo Castle.[8] The warriors fighting on the Iga side in *Iran ki* are usually called simply Iga samurai 伊賀侍, *jizamurai, kunimusha* 国武者 (provincial warriors) or *gōshi* 郷士 (country samurai).[9] In one section Kikuoka writes simply about 'ten men who had shinobi skills 忍びの術を得た',[10] and the now familiar expression 'shinobi no mono' appears only once in the entire work.[11] As a general rule, when 'shinobi' is used in *Iran ki* it is an adverb describing things done in secret by troops of both sides, as when Iga's enemy Takigawa Kazumasu sends troops secretly into the castle of Maruyama, which he is hastily repairing.[12]

The First Iga Rebellion

The first Tenshō Iga no Ran may be traced back to 1571 when Nobunaga, who by then firmly controlled the Shogun Ashikaga Yoshiaki, used the Shogun's prerogative to appoint Nikki Yoshimi as *shugo* of Iga. There appears to have been no hostile reaction from the Iga Confederacy to Nikki's appointment, which may have been related to the fact that Iga and Kōka had recently lost nearly 800 of their best men in the battle of Yasugawa, but Nobunaga's attentions had anyway shifted to the province of Ise. Ise was of much greater strategic importance to Nobunaga than little landlocked Iga. It had a long coastline, the great road of the Tōkaidō ran through much of it, and it had a common border with Oda Nobunaga's home province of Owari.

Ise had seen its own local version of the transition from *shugo* to *daimyō* in the shape of the Kitabatake family, a house of glorious pedigree. In 1569 Kitabatake Tomonori (1528–1576) revolted against his elder brother, an appalling act of disloyalty that gave Nobunaga a useful pretext to attack their castle of Ōkawachi.[13] He began by burning to the ground the town around it and sealing off the fortress using bamboo palisades. The battle was a fierce one and twenty of Nobunaga's high-ranking samurai were killed, along with about 300 other ranks.[14] Greatly frustrated, Nobunaga ordered a scorched earth policy for the surrounding area with the aim of starving the defenders to death, but when the siege had lasted fifty days Kitabatake proposed peace terms that included his adoption as heir of Nobunaga's second son. This was agreed and the eleven-year-old child was named Oda Nobukatsu (1558–1630). Two guardians were appointed, both of whom had their origins in Iga. They were Nobunaga's second-in-command Takigawa Kazumasu (1525–1586) and Tsuge Saburōzaemon (?–1579).

Kitabatake Tomonori continued to rule Ise as a puppet *daimyō* in the sad knowledge that when he died his inheritance would pass to Oda Nobukatsu, the cuckoo in the Kitabatake nest. That moment came in 1576 and it was widely believed in Ise that Tomonori was murdered by Tsuge Saburōzaemon. Oda Nobukatsu consequently inherited Ise, but the surviving members of the Kitabatake family did not accept his succession meekly. Takigawa Kazumasu put down the rebellion with great severity and many of the defeated Kitabatake samurai fled to Iga. From there they appealed for help to Nobunaga's deadliest enemy, Mōri Terumoto, whose territories lay to the west of Kyoto and who was a supporter of the Ikkō-ikki. At about the same time the Iga Confederacy challenged Nobunaga in a different way by ousting their *shugo* Nikki Yoshimi.[15]

Unfortunately for the Iga Confederacy, their solidarity was not complete and in 1579 a disaffected *kokujin* called Shimoyama Kai-no-Kami visited Oda Nobukatsu to complain that his fellow countrymen 'were living in luxurious excess and were not governed by the dark truths of Heaven'.[16] This gave Nobukatsu the pretext for chastising them, and he recalled that some years previously Kitabatake Tomonori, planning one day to conquer the province himself, had built a castle almost at the centre of Iga Province on a hill called Maruyama. Nobukatsu realised that such a base would be essential to his own operations, so he ordered Takigawa Kazumasu to restore it. The Iga *ikki* decided to launch an attack before Maruyama was finished and set up headquarters in another abandoned castle site across the river called Tendōzan, which is now the site of a temple called Muryōjufukuji. Takigawa tried to pull his army together after their surprise attack, but because of his incomplete defences he was forced to withdraw to the nearby village. Here his men were assaulted by more Iga samurai, and Kikuoka delights in telling us that one fought on even though he had his left arm cut off.[17] The Ise troops who were still in Maruyama Castle eventually fled because they knew that their attackers 'possessed techniques for entering castles'.[18] The Iga warriors pursued the Takigawa samurai into wooded valleys and flooded rice fields. Spies (written as shinobi no mono for the only time in *Iran ki*) reported that Takigawa had escaped back to Matsugashima, and regret was expressed at their failure to kill him.[19]

65

The angry Oda Nobukatsu sought revenge and planned a three-pronged attack on Iga from Ise using the three main mountain passes, but more spies (here written as *kanchō*) from Iga once again did their work well, so the men of Iga were able to make preparations.[20] Nobukatsu set off in grand style, and here Kikuoka waxes lyrical as he describes how 'ten thousand banners fluttered in the autumn breeze, and the sun's rays were reflected off the colours of armour and *sashimono*[21] ...his gold umbrella standard came out of the black cloud of the thick morning fog from which he advanced, as wonderful as the rising of the sun'.[22] The Iga samurai were ready for them, and attacked the Ise army in a classic ambuscade:

> The Ise samurai were confused in the gloom and dispersed in all directions. They ran and were cut down in the secluded valley or on the steep rocks. They chased them into the muddy rice fields and surrounded them... Some killed each other by mistake. Others committed suicide and it is not known how many thousands were killed.[23]

The army of Tsuge Saburōzaemon entered Iga by a different pass, where 'he was pursued and surrounded by spears. Several hundred soldiers flocked round him to take their vengeance, all stabbing him together. They stabbed him many times until he died'.[24] The death of a general on the battlefield at the hands of a mob was an unusual and disgraceful occurrence and *Shinchō-Kō ki* mentions it in the two brief sentences it devotes to the campaign. 'On the 17th of the 9th month, Kitabatake Chujō Nobukatsu crossed with his forces into Iga Province on a punitive expedition. A battle flared up and Tsuge Saburōzaemon was killed'.[25] His death is recalled later in the harsh words of Nobunaga's letter to his son:

> The other day you were guilty of a fiasco on the Iga border. Take this as a lesson: The Way of Heaven is terrible indeed, and the sun and moon have not yet fallen to the earth. Why did you do what you did? ...On top of everything else you sent Saburōzaemon and the others to their deaths. That is unspeakable...[26]

The Second Iga Rebellion
The year 1580 found Oda Nobunaga preoccupied with the Ikkō-ikki, whose ten-year-long campaign against him was finally brought to a

peaceful conclusion. Meanwhile the Iga Confederacy kept up their vigilance, including the raid outside the province on the fortress of Sakaibe, which was described earlier. By 1581 Nobunaga was ready to turn his attentions towards Iga once again. This time Nobunaga himself took the initiative, although he did not lead the army in person, and once again treachery within the ranks of the Iga Confederacy aided his arrangements because, 'Nobunaga had pardoned Fukuchi of Tsuge, and hostages were taken to secure his loyalty,' and 'A man called Taya of Kawai surrendered to Nobunaga, pleading for mercy'.[27] Nobunaga planned an invasion by his generals over six different synchronised attack routes. A detailed army list appears in *Shinchō-Kō ki*.[28] It includes a contingent from the newly pacified Kōka.

The Iga *ikki*, whose intelligence service was as reliable as ever, knew that they did not have the resources to ambush six separate armies each the size of the total they had defeated in 1579. They consequently assembled their main forces in two places that were suitable for quick dispersal: the Heirakuji, a temple on a hill in the middle of Ueno village, which is now the site of Iga-Ueno Castle, and Tendōzan, the site of the attack on Maruyama Castle.[29] The account of Nobunaga's army crossing the border shows a difference in tactics from 1579 on his side as well. Their plan was to destroy everything and everyone, so 'he raided the village on the Ise Road and burned people's houses to the ground', says *Iran ki*.[30] The tactical process is confirmed in *Shinchō-Kō ki*'s description of the army 'burning to the ground completely the temples and shrines within the province, beginning with the shrine precincts of the Ichinomiya'.[31] An *ichinomiya* was the principle Shinto shrine in a province, which in Iga's case was the Aekuni shrine. This policy was adopted on all six routes, with the overall aim of forcing the Iga defenders to abandon their villages and take to fortified positions, thus reducing the opportunity for the guerrilla tactics that had worked so well in 1579. Elsewhere in Iga, 'one hundred heads were placed in a line of the brave samurai from nearby villages who fought against the enemy', while some desperate Iga samurai 'put to the sword their ten children and their wives, and set off with light hearts to be killed in action, knowing that their wives and children would have been captured alive and carried off to other provinces'.[32] The devastation of Iga and its effects on 'men and women, the old and the young' even reached the ears of the author of *Tamon'In nikki*, who comments sadly on the news.[33] In the Hijiki area Nobunaga's army burned to the ground a Buddhist foundation called Kinsenji:

This sad and sacred place which stood there was scoured, and only rubbish adorned it. When the smoke died down, inside and outside were dyed red with blood. The corpses of priests and laymen were piled high in the courtyard or lay scattered like strange autumn leaves lying deep of a morning.[34]

One of the new ways into Iga in 1581 was the Tamataki route from Kōka, and in command was Gamo Ujisato (1556–1599). The fighting there was particularly savage, with 'the wives and children of our families fleeing hither and thither from this place to that place, but because of the attack they were cruelly slaughtered, mown down like blades of grass'.[35] It was in this sector of the campaign that the attack was made against the fortress of Amagoisan, noted above for the regulations in which shinobi are mentioned, and the use of secret operations by the Iga side is also related by Kikuoka in *Iga Kyūkō*. 'On the night of the 28th, many experts in shinobi no jutsu (*ninjutsu no tassha* 忍術ノ達者) from the province secretly entered Ujisato's lodgings but were driven off by the guards who were on duty'.[36]

Iran ki then devotes several pages to a major attack on Hijiyama Castle, the site of which lies across the flat rice fields to the west of modern Iga City. Gamo Ujisato began burning the villages which are now the outlying suburbs of Iga. Tsutsui Junkei (once the ally of Iga and now a senior vassal of Nobunaga) joined him for the attack, setting up a base on nearby Nagaokayama. The defenders tried to ambush the attackers during an assault on the main gate:

> The troops in ambush pushed into them and ran round killing...
> They shot and thrust, and threw great rocks and great trees from the edge of the ditch. They attacked them with guns fired from loopholes from a distance. The enemy who managed to approach were greatly disconcerted, and many were exhausted. The majority were wounded, and many were lying on the ground.[37]

Following this tremendous effort the garrison held a council of war, preceded by the traditional ceremony of naming their seven best fighters as the 'Seven Spears of Hijiyama'.[38] They were obviously not content to rest on their laurels, and Kikuoka's heroic narrative continues with an exhortation to use the undercover skills for which their province would become famous. 'Let us risk a night attack on Nagaokayama and take Junkei's head, which will be amazing to the eyes of the enemy and will add to the glory of the province'.[39] They launched the attack by the light of pine torches on the 1st day of the 10th month (28 October 1581), but raiding could not keep the siege at bay for ever. Food was running low and *Iran ki* describes how 30,000 Oda troops surrounded the mountain and raised their war cry as they prepared for the final assault on Hijiyama. The shout was met by silence from the grim defenders, every man of whom 'looked like a wooden statue of the Buddha' as he stood motionless waiting for the attack.[40] The weather was dry and a strong wind made the conditions ideal for the most deadly weapon in the samurai armoury. 'They set fire to all the temples over a wide area. This time there was no rain to be blown by the wind. The flames blazed and spread as a sign to the whole world. Some fires were extinguished, but it was many months before the black ashes died down'.[41] With Hijiyama destroyed Nobunaga's generals were given a free hand to scourge the area, demolishing forts as they went.[42]

Nobunaga's armies then joined forces for an attack on a castle in the south of Iga Province, which would prove to be the final battle of the campaign. Kashiwara Castle was under the command of Takino Jurō Yoshimasa, and its spirited defence allows Kikuoka to write some of his liveliest passages.[43] In all 418 samurai and 1,200 foot soldiers, together with an unknown number of women and children, were crammed into the mountain fortress. It was a strong place because, 'the fortress of Kashiwara Castle had been constructed in a makeshift manner, but on its eastern side was a tall cliff and a long forest so that arrows loosed or stones thrown at it could not reach further than this shield of trees, and penetrating from the outside was impossible'.[44] Two days after the fall of Hijiyama, Gamo Ujisato, Niwa Nagahide, Takigawa Kazumasu and Tsutsui Junkei joined Oda Nobukatsu with 7,000 fighting men in all.[45] An initial attack was beaten off with heavy losses, so Oda Nobukatsu decided to starve the defenders out. Food soon ran low, so the garrison tried a stratagem which, even if it did not persuade the besiegers to depart, might allow the opportunity for some of the non-combatants to escape. The plan was for one or two soldiers to slip out (*shinobi dete* 忍 び出て) through the enemy lines and contact the nearby villagers, who would prepare hundreds of pine torches to make Nobukatsu think that a relieving army had arrived. Three men disguised themselves as beggars and in the middle of the night of 12 November hundreds of lights began appearing outside the siege lines, causing great alarm. It was Niwa Nagahide who calmed his army's fears. He knew his country's history and recognised that this was a stratagem used in the fourteenth century by Kusunoki Masashige. The ploy therefore failed, but the defenders had other tricks up their sleeves:

> However, the enemy became increasingly careless, including many of the guard outside the castle, and ten Iga-mono who had mastered shinobi no jutsu 忍びの術 lay in wait in various places outside the castle as the light faded and it grew dark. At night they slipped out in secret and made night raids on the camps of all the generals and set fire to them using various techniques. Many of the enemy were thus plagued, making them no longer careless. They then defended better against the incursions. However, Niwa Nagahide's camp was attacked by night on several occasions, and night after night his followers were killed. Over a hundred men were slaughtered, and because of this the enemy were highly scared and lost their vigour, realising that they had no sense of security.[46]

In spite of these heroics the demoralised and starving garrison were reduced to planning a suicide attack for 21 November, which would either break the enemy or endow the defenders of Kashiwara with glorious names for posterity, but the conclusion of the siege of Kashiwara came about not as a result of a battle, but by peace negotiations. This remarkable development arose through a visit from a man from Nara called Ōkura Gorōji. He is described as a teacher and performer of sacred dances at Shinto shrines, and his mediation was enlisted by Oda Nobukatsu. After meeting both sides and gaining Nobukatsu's assurance that he would spare the lives of the garrison, Ōkura Gorōji arranged for Takino Yoshimasa to visit the enemy camp, where Takino presented Nobukatsu with a diplomatic gift of a saddle and handed over his son as a hostage for good behaviour. Takino received in return five gold pieces and a black horse, having suffered a tirade from Nobukatsu, which Kikuoka expresses as, 'The people of Iga are conceited, holding to personal opinions, they are like frogs in a well who know nothing of the great ocean. They despised their *shugo*, wanting their own way, not observing due proprieties, not submitting to the exalted person of the Son of Heaven', etc., etc.

On 24 November the castle was evacuated and occupied temporarily by Tsutsui Junkei,[47] but in the north of Iga Province a strange drama had been played out while the siege of Kashiwara was still in progress. Because of the fall of Hijiyama Oda Nobunaga was finally able to go and see for himself the province that had caused him so much trouble, and on 6 November he arrived at the site of the Aekuni shrine that his troops had burned to the ground. His generals erected a temporary mansion on a nearby hill, where he stayed for three days.[48] The most interesting point about the above section is what it does not include, because *Shinchō-Kō ki* diplomatically omits an attempt on Nobunaga's life that is supposed to have happened during his visit. Kikuoka, naturally enough, recounts it with pride, but the fact that it is included out of sequence suggests that it is a story that has grown in the telling. Three Iga samurai lay in wait for the hated Nobunaga who had brought them so much misery, and as Nobunaga sat surrounded by his followers they fired guns at him from three different directions. Although the shots missed Nobunaga they succeeded in killing seven or eight of his retainers. Kikuoka concludes this extraordinary anecdote by a reference to his homeland's great tradition, 'The one among them called Otowa no Kido ('Kido of Otowa') had acquired the mysteries of *bōjutsu hijutsu* 謀術 火術 ('techniques of deception and fire')... The Iga *shinobi* 伊賀忍 of today follow in the tradition.[49] My understanding of the expression 'techniques of deception and fire' means that Kido was an excellent sniper. It would also fit with Kikuoka's comparison with the Iga-mono of Edo Castle, because part of the role of the palace guards was to act as a musketeer squad.

The settlement of Iga
The terrible Tenshō Iga no Ran therefore ended peacefully, even though many people had already been slaughtered. Yokoyama Takaharu has calculated that 4,000 men and women died,[50] but instead of carrying out a final punitive massacre Nobunaga behaved with mercy towards the survivors, even if all their temples, shrines, castles and mansions were burned to the ground along with any historical records they may have contained. The *ikki* members who had not already fled to other provinces were quickly disarmed as the province passed rapidly under the new and unwelcome situation of *daimyō* control. Three out of Iga's four districts were assigned to the victorious Oda Nobukatsu and the remaining one was given to Nobunaga's brother Oda Nobukane (1548–1614), so from 1581 until the time when Kikuoka was writing and long beyond that, Iga Province was a *daimyō*'s domain.[51]

In 1582, however, a brief and little-known flurry of activity saw the Iga samurai behaving quite like their old selves in a minor campaign sometimes referred to as the Third Tenshō Iga no Ran. The background was the sudden and totally unexpected death of Oda Nobunaga. He had been resting at the Honnōji in Kyoto, making ready to move west to help in the fight against the Mōri. Realising that Nobunaga was unusually isolated and weakly defended, a treacherous general called Akechi Mitsuhide launched a surprise attack. Completely overwhelmed, Oda Nobunaga committed suicide. Mitsuhide's men then sought out Nobunaga's heir Nobutada and killed him too. The news quickly spread to Iga and while some castle commanders stayed loyal to the apparently doomed Oda family, others saw Akechi Mitsuhide's coup as the opportunity to revisit the glory days of the Iga Confederacy. One of these opportunists was Morita Jōun, who was killed trying to take over Ichinomiya Castle for himself.[52]

At about the same time there occurred the celebrated story of the escape of Tokugawa Ieyasu through Iga. The usurper Akechi Mitsuhide had moved rapidly to eliminate Nobunaga's family and seek out his closest allies, including Ieyasu, who was trapped in Sakai with a small defensive retinue. His direct route back to Okazaki would have taken him through Kyoto, which was under Akechi's control, so Ieyasu risked a devious route through Iga on the advice of his loyal retainer Hattori Hanzō, a man born in Mikawa to an old Iga family. Ieyasu's epic journey is recounted with pride in Iga and is one of the reasons why the name of Hattori Hanzō has risen to such prominence in the ninja myth. Ieyasu's retainer Anayama Baisetsu did not accompany him on the Iga route and was ambushed and killed on the way, so the peril was genuine. The 1833 *Mikawa Gofudo ki* is the main source for the colourful story, from which the following extract is taken:

> Yamaoka and Hattori accompanied them, defying mountain bandits and *yamabushi* alike... Hattori Sadanobu was praised for the great extent of his loyalty, and on leaving he was presented with a *wakizashi* forged by Kunitsugu. ...Hattori Hanzō Masashige was an Iga man. Sent on by Tadakatsu, he went ahead as guide to the roads of Iga. The previous year, when Lord Oda had persecuted Iga, he had ordered, 'The samurai of the province must all be killed'. Because of this people fled to the Tokugawa territories of Mikawa and Totomi, where it was ordered that they be shown kindness and consideration. Consequently their relatives were able to pay them back for this kindness... They also rid of a number of notorious bandits.[53]

Tokugawa Ieyasu never forgot the debt he owed to the people of Iga. In 1590, having served with distinction in the Odawara Campaign, Ieyasu was granted the castle of Edo, which is now the Imperial Palace in Tokyo. In a moment of inspiration Ieyasu chose loyal Iga men to serve him as a hereditary guard unit in the new fortress and Hattori Hanzō was placed in charge of them. The reward was a great honour and would prove to be a very important addition to the building blocks of the ninja myth.

In contrast to the situation of the Iga men within Edo Castle, an unsettled atmosphere persisted for many years within the province itself in spite of strict *daimyō* control. There were a number of agrarian riots, one of which was mounted against the Tōdaiji, whose monks were still landowners in spite of the war.[54] The belief that the Iga shinobi tradition not only survived, but even prospered under these circumstances is a vital part of the ninja myth, which states that Iga shinobi who were desperate for employment marketed their unique skills and began to sell their services more widely than ever before, fighting for other *daimyō* as mercenaries and setting up schools of ninjutsu.[55] There is no reason to rule out some form of migration at the very least, and there does appear to be the authentic example from *Arayama kassen ki* quoted earlier of Iga 'robbers' fighting for the *daimyō* Maeda Toshiie in 1582. This is one of the earliest examples from the Tokugawa Period of an undercover operation being linked directly to men from Iga and is also probably the first to mention them being present at any battle that took place after 1581.

The authenticity of the idea that 'Iga robbers' fled north after Nobunaga's invasion is also supported indirectly by the presence in Taikō ki of another reference to Iga in the context of the capture of the castle of Chungju during the invasion of Korea in 1592. In Yoshida's 1979 re-telling of the story we read that the Japanese commander Konishi Yukinaga had serving under him 'one hundred Iga no shinobi no mono'.[56] However, in Kuwata's earlier version from 1943 where the language is closer to the original the phrase he employs refers to 'one hundred Iga-mono (used) as Konishi's shinobi no mono', which suggests that the original author is simply showing his familiarity with the Iga-mono stationed in Edo Castle by using that name for whatever undercover troops Konishi Yukinaga may have had with him.[57]

This passage is almost certainly an example of Oze Hoan's tendency towards exaggeration that is typical of the Tokugawa Period, in this case using what he knew about the Iga-mono in Edo. The earlier section lifted from *Arayama kassen ki* is different because it was not subject to Oze's editing. *Arayama kassen ki* therefore supports both the dissemination hypothesis and Iga's claim to a particular hereditary expertise at undercover work exercised through lower-class warriors who lived outside the law. It also has added significance because of Oze Hoan's juxtaposition of this authentic account of Iga shinobi in action with the fictional one of them fighting in Korea. *Taikō ki* may therefore mark the precise moment of the start of what would become the Iga ninja myth with Oze Hoan as its first inventor of tradition.

Apart from this apparently unique instance, Nobunaga's invasion of Iga did not mean the start of an authentic Iga shinobi tradition; it meant the end of it, and as reality faded imagination began to take over to produce the ninja myth of today. The sequence of events began when Oda Nobunaga died at the Honnōji in 1582. With his death chaos returned to Kyoto, but the disorder lasted for only a short time because peace was restored within days by his general Toyotomi Hideyoshi. Hideyoshi hastily patched up a peace settlement with the Mōri and rushed back to Kyoto, where he defeated the usurper Akechi Mitsuhide at the battle of Yamazaki. Oda Nobukatsu, the veteran of Iga, expected to succeed his father, but Hideyoshi cleverly outflanked him by declaring that Nobunaga's infant grandson by his dead elder son would be the new heir. His uncle Oda Nobukatsu was dispossessed and humiliated. Smarting under the insult, Nobukatsu made the fateful decision to fight against Hideyoshi but was defeated at the battle of Nagakute in 1584 and deprived of his lands. Using the pattern of resettlement that Nobunaga had begun Hideyoshi replaced recalcitrant *daimyō* with his most loyal generals, so in 1584 Tsutsui Sadatsugu (1562–1625), the cousin and adopted heir of the recently deceased Tsutsui Junkei, received Iga. In Lamers' words, the Tsutsui were 'unknown and unloved' in Iga, but that statement is only half true.[58] As three generations of the Tsutsui had fought with or against the Iga *ikki* for half a century and taken part in the destruction of Iga in 1581 they were far from being unknown to their new subjects. They are, however, highly likely to have been unloved.

Tsutsui Sadatsugu would lead his new subjects from Iga into battle for the first time in 1585. The occasion was Hideyoshi's campaign against Kii Province (modern Wakayama Prefecture). It must have been an interesting experience for the Iga men because they were taking on a confederacy reminiscent of their former status. The Kii *ikki* consisted of a formidable mixture of *jizamurai*, *kokujin* and other warriors whose religious affiliations gave them a common cause. Among them were the Negoro-shū, from the militant Shingon temple of Negorodera, and the

Saika-shū, the last remnants of the Ikko-ikki, whose main base in Osaka had been pacified by Nobunaga in 1580.[59]

The Kii campaign began when Hideyoshi sent a peace envoy to Negorodera. The man was subjected to a gunfire attack on his quarters and rapidly fled back to Kyoto. Hideyoshi then launched an invasion using a fleet of ships to attack the *ikki* by sea as well as land, concentrating on four outlying forts. Tsutsui Tadatsugu and his Iga-shū were involved in a desperate attack on the fort of Sengokubori on 20 April, which ended with the severing of the heads of the entire garrison.[60] Some extra details are added by *Taikō ki*, which has Tadatsugu leading an assault by Hideyoshi's Horse Guards, but even though they killed over 300 defenders they were driven back by fierce gunfire. Someone within the Tsutsui contingent then loosed a fire arrow into the castle and started a conflagration that spread to its gunpowder store. As *Taikō ki* puts it, 'in the blink of an eye, along with a great noise like a thousand thunderbolts, the castle was reduced to ashes, and over 1,600 renowned warriors of Kii Province were together destroyed in the flames'.[61] Taking advantage of the sudden devastation, Hashiba Hidetsugu led a successful assault and Sengokubori fell. All this is perfectly believable, but it is interesting to note that nothing more was ever made of the fact that it must have been a member of the Iga-shū who was responsible for the fall of the castle by his skilled archery: a 'ninja operation' if there ever was one!

Following the battle of Sengokubori the warriors of Iga Province seem to fade rapidly from view. During the siege of Ichinomiya on Shikoku Island in 1585 Tsutsui Sadatsugu's Iga-shū dug a secret tunnel into the castle and Tadatsugu's service was commended in a letter from Hideyoshi, but little else is known about their military activities around this time.[62] In 1590, of course, Hattori Hanzō had taken charge of the prestigious unit of Iga guards in Edo Castle, but no comparable service to the Tokugawa was ever performed back in Iga Province itself. By contrast, when the great Sekigahara Campaign began in 1600 Tsutsui Sadatsugu was so disinclined to lead his Iga warriors into battle on behalf of Ieyasu that neither samurai nor shinobi from Iga played any part in that decisive struggle. Instead Sadatsugu simply locked himself inside Iga-Ueno Castle and sat out the war. That was how he lost Iga, because his failure to provide active support to the victor Tokugawa Ieyasu resulted in the confiscation of his lands in 1608.

As part of Ieyasu's widespread redistribution of territories, Tōdō Takatora (1556–1630) was moved from Imabari on Shikoku Island to the fief of Anotsu, which included Ise and Iga Province.[63] His descendants would reside there peacefully and uneventfully until 1868, and the quiet castle town of Iga-Ueno became famous not because of ninja, but for being the birthplace of Matsuo Bashō, Japan's greatest poet. An early tourist guide to Japan of 1891 merely notes that Ueno is 'some 23 miles from Tsu'.[64] The Iga component of the ninja myth was lying dormant and would slumber on like Sleeping Beauty, ready to be awakened during the twentieth century by a kiss from a handsome prince.

Chapter 6

The Shaking of Kōka

With Iga Province pacified and samurai from Iga acting as élite guardsmen in Edo Castle, the ninja myth could be advanced only by men from Kōka, whose secret operations added considerably to its building blocks over the ensuing three decades, although not in the way that one might expect. It all began when the Kōka-shū passed automatically under the control of Nobunaga's successor Toyotomi Hideyoshi and fought for him during the Kii campaign. The successful reduction of the Kii forts in which Iga had cooperated had been followed by the surrender of Negorodera, so Hideyoshi was able to concentrate totally on Ōta Castle. On 24 April Nakamura Kazuuji was sent as an envoy along with Suzuki Shigehide to urge the garrison to surrender, but they refused. Ōta lay near the Kii River, so Hideyoshi constructed a long encircling earthen rampart combined with a dam to divert its waters and flood the area.[1] The speed of construction is probably exaggerated, but all the accounts report that by 30 April water was beginning to enter the castle, and over the next two days its level grew rapidly, helped by exceptionally heavy rain.[2]

The circumstances of the flooding of Ōta allowed many opportunities for individual daring, and it is interesting to see the word shinobi appearing in its adverbial form in the accounts. In *Ōta mizuzeme ki*, 'In the middle of the night of the 5th day of the 4th month, a man from the attacking force called Amagasaki Kichibei, sailing on a boat, secretly entered the castle, but he was seen by the men in the watchtowers who made him turn back. Lord Hideyoshi graciously heard this, which put him in a good mood'.[3] Similar accounts in other sources relate how the man was soon killed, and one tells us that Amagasaki Kichibei was actually a pirate:

The pirate (*kaizoku*) Amagasaki Kichibei with one hundred men under his command and attired brilliantly in his armour, sailed up near to the castle and delivered an insult in a loud voice, at which the castle garrison made a plan to entice him. The castle commander Ōta Gendaiyū took a musket and fired it from a loophole, the bullet hit Kichibei right on target and he fell to the bottom of the boat.[4]

A few sentences later it is reported that the flooded castle was becoming a haven for 'rats, weasels, otters and snakes'.[5] Another man entered secretly to gain intelligence on the night of the 7th day to assess the state of the garrison.[6] The floodwaters were now so deep that Hideyoshi was able to sail large vessels in from the sea and bombard the castle, but the defenders hit back when skilled swimmers among the garrison dived under the ships, made holes in their hulls and sank them.[7] Nevertheless, a message eventually arrived from the beleaguered castle suggesting that they would surrender if Hideyoshi spared their lives. One hundred and fifty of their best warriors would commit suicide as a gesture of submission. This was agreed, and Nakamura Kazuuji made a triumphant entry.[8]

The Ōta Campaign therefore provides further evidence that shinobi activity could take place when no one from Iga was anywhere near, so it is very interesting to read that the Kōka-shū were present within Hideyoshi's army at Ōta. Their role during the siege was to have a major impact upon the future of both Kōka and the ninja myth, even though their contribution to the battle was almost a disaster. This embarrassing situation arose because different units of Hideyoshi's army had been placed in charge of separate sections of the dyke. The sector that was the responsibility of the Kōka-shū collapsed, probably because the enterprising garrison had raided the area and created the breach.[9] The rushing waters flooded the camp of Ukita Hideie and caused many casualties.[10] Repairs were undertaken and the overall operation was of course a success, but the Kōka-shū would be severely punished for their negligence, because within one month the Kōka District had been handed over to Nakamura Kazuuji.[11]

The transfer happened on 5 June 1585 and was a fateful date in Kōka's history.[12] The move was perfectly in accord with Hideyoshi's long-standing policy of transferring *daimyō* around the provinces and was a suitable reward for Nakamura Kazuuji, so the imposition of a new ruler was not in itself a punishment for allowing the dyke to be breached, but it meant that the new *daimyō* would be the one to carry out a suitable chastisement. Nakamura Kazuuji began by consolidating his newly created fief of Minakuchi in the manner that was now common by the creation of one formidable castle to replace a number of smaller ones. He chose a prominent hog's-back mountain overlooking the Tōkaidō at Minakuchi. The new fortress was named Minakuchi Okayama Castle.[13] Along with the lands Nakamura inherited the Kōka-shū, within whose ranks twenty families were due for punishment because of their negligence at Ōta. They therefore suffered *kaieki* 改易 (attainder), the process of having their small landholdings confiscated and losing the status of samurai. A later account of the process tells of 'the samurai-shū becoming *rōnin*' and thereby losing ownership of 'dry and wet fields, mountains and forests'. The drastic treatment of them became known as the *Kōka Yure* 甲賀ゆれ, the 'shaking of Kōka', and the 'shaking' was not confined to former *jizamurai* becoming farmers.[14] Many temple and shrine lands also passed into the hands of Nakamura Kazuuji. The situation is summarised in the words of *Yagawa Zakki*, the records of the shrine of Yagawa Daimyōjin in Kōka:

> ...after twenty-one families had received punishment by attainder from Lord Hideyoshi and were placed in confinement, even priests fled. In a similar manner to that of the twenty-one families the temple lands of the Yagawadera were also confiscated... If we listen to the tradition, at the time when Okayama Castle was built, there is no disagreement that to supply materials for the building of the castle they made off with good timber from temple buildings, also old manuscripts, even down to the foundation stones. Was there such an urgent need?[15]

The 'Shaking of Kōka' further illustrates the point that independence of military action was impossible under *daimyō* control unless it took the form of an armed rebellion, and for that to happen a potential rebel would need weapons. To counter that possibility Toyotomi Hideyoshi set in motion his notorious *katanagari* 刀狩り (Sword Hunt). This was the operation by which Hideyoshi's agents forcibly confiscated all weapons from anyone except a *daimyō*'s samurai, most of whom were now completely separated from the land, based in castles and in a state of vassalage to Hideyoshi's loyal generals. Even the inhabitants of the holy mountain of Kōyasan were ordered to disarm.[16] The Sword Hunt was so thorough that it would be incredible to think that any warriors still had the means to fight as mercenary shinobi as the ninja myth suggests.

The Kōka-shū in the Sekigahara Campaign

Toyotomi Hideyoshi died in 1598 leaving his five year-old son Hideyori to inherit the newly unified realm of Japan. Perceiving the weakness of a child ruler, two armed factions developed under Ishida Mitsunari and Tokugawa Ieyasu. Their rivalry came to a climax at the decisive battle of Sekigahara in 1600, which resulted in a victory for Ieyasu and the founding of the Tokugawa Shogunate. Kōka's involvement in the Sekigahara Campaign is a complicated story because a certain amount of dispersal had taken place when they were defeated by Nobunaga and there was also a long tradition of service to the Tokugawa. In 1597 Ieyasu granted a stipend to Takamine Shinzaemon and ten other Kōla men,[17] while in Northern Ōmi men of Kōka served under Ieyasu's general Yamaoka Dōami Kagetomo (1540–1603), who had been entrusted by the Tokugawa with the castle of Nagashima.[18]

As a long-standing Tokugawa loyalist Yamaoka Dōami had rendered great service to the Oda and Tokugawa at the time of Nobunaga's death in 1582. He was at that time the keeper of Seta Castle beside the strategic Seta Bridge, which he destroyed in an attempt to prevent the usurper Akechi Mitsuhide marching against Azuchi Castle. The ploy succeeded only in delaying Akechi's destruction of the great fortress, but Dōami also made a contribution to the Tokugawa cause when he took part in assisting Ieyasu's escape through Iga. In 1600 Dōami's Kōka-shū were in Nagashima when the Sekigahara Campaign started, although their homeland of Kōka had become enemy territory because the fief of Minakuchi had passed from Nakamura Kazuuji to Natsuka Masaie (1562–1600). Masaie eventually declared his support for the rival faction under Ishida Mitsunari at the very last moment. It was a decision that would cost everyone dear.

The great Sekigahara Campaign began when Tokugawa Ieyasu was forced to move north to deal with a threat from Uesugi Kagekatsu in Tōhoku. As Ishida Mitsunari was active around Kyoto it was very important to both rivals that their enemy's castles should be captured, and one of these places was Takatori in Yamato Province. A man from Kōka who 'had shinobi skills' was chosen to sneak into Takatori by climbing up crags and tree roots. Anticipating just such a move, the garrison had set up trip ropes attached to wooden clappers and the man was captured alive. They sliced off his nose and ears and cut off his fingers and toes. He was then sent back to the besiegers as an example of the treatment they might expect.[19]

The most strategic Tokugawa base was Fushimi to the south of Kyoto, which had men from Kōka actively engaged in its defence in one of the most decisive actions of the Sekigahara campaign. Fushimi was under the overall control of Torii Mototada and among the garrison were over a hundred samurai from Kōka, sent to Fushimi under Yamaoka Dōami's younger brother Kagemitsu. They were stationed in the castle bailey known as the Nagoya-maru. No impression was made upon the fiercely defended fortress, so a cunning plot was laid. Natsuka Masaie of the Ishida faction was lord of the castle of Minakuchi in Kōka where the families of the Kōka men resided, so he seized some hostages in the absence of their menfolk. An arrow letter was loosed into the castle, informing the men (whose numbers vary from two to eighteen depending upon the source) that if they cooperated by setting fire to Fushimi they would be richly rewarded. If they refused their wives and children would be crucified. In order to save them the Kōka men betrayed their comrades, set fire to a tower and took down a section of the wall.[20] The enemy broke in and, after much desperate fighting, Fushimi fell with a huge loss of life including those of the Kōka contingent who had remained loyal. They had, however, inflicted considerable losses upon their enemies and bought precious time that Ieyasu was able to exploit at the battle of Sekigahara. The traitors from Kōka were captured alive and then crucified on Ieyasu's orders.[21]

After Sekigahara Yamaoka Dōami took terrible revenge when he defeated Natsuka Masaie in Minakuchi Castle. Having fled there from Sekigahara without firing a shot, Masaie knew his cause was hopeless and committed suicide. The Minakuchi fief then passed under the direct control of the Tokugawa family while Ieyasu considered Kōka's position. He concluded that long service rendered to the Tokugawa by Yamaoka Dōami and the death of a hundred men from Kōka outweighed the act of treachery by a few, so the descendants of those who died at Fushimi were taken into Ieyasu's service and Dōami was placed in charge of them. He was granted 9,000 *koku*, 4,000 of which was for his Kōka estates,[22] and the next we hear of the Kōka men is that they are to be found serving as guards in Edo Castle alongside the Iga men in a unit consisting of ten mounted samurai and a hundred foot soldiers. These same men went on to serve at the siege of Osaka in 1614–15, but in the capacity of a troop of musketeers, not shinobi. Together with similar units from Iga and Negoro, these *hyakunin-gumi* (hundred-man squads) defended Edo Castle using the finest modern weapons then available.

The Kōka warriors' appointment to Edo was a reward for good service, just as it had been for their former Iga comrades who had helped Ieyasu's escape in 1582, although both gestures are conventionally explained as a conscious act whereby Tokugawa Ieyasu took the supposedly powerful and independent military cults of the Iga and Kōka ninja under his own wing. He then gained exclusive use of them and thus effectively neutralized any capability a possible rival might have to recruit shinobi. I now believe that the situation was quite different, because their appointment as guards at Edo Castle shows how firmly the Iga and Kōka-shū had been integrated into Tokugawa military society rather than ever having been separated from it. This aspect of the ninja myth therefore derives from their close relationship with the ruling Tokugawa family and not from any notional independence of action.

The Kōka petitions
The sons and grandsons of the loyal Kōka-shū from Fushimi entered Edo Castle as heroes whose appointment had placed them at the very heart of the Tokugawa hegemony, but back in Ōmi Province lived the twenty families who had suffered attainder under Hideyoshi and had been reduced to the status of farmers. In time the policies of the Tokugawa Shoguns would set in stone the distinction between samurai and workers of the land, and when the Kōka Koshi 甲賀古士 ('The Old Kōka samurai') began to suffer hardship during the 1660s their leaders protested. They did not riot; instead a series of sympathetic local officials petitioned the Shogun over a period of three decades for a restoration of their samurai status. In 1667 Akutagawa Jingōbei Toshishige presented the first petition to the Shogun.[23] The document bears the title of *Osore nagara sojō wo motte gonjō tsukamarisōrō* ('a petition... served most humbly'). It consists of seven sections, and in each part the Kōka Koshi's ancestors' service to the Tokugawa family is delineated. These accounts would grow in the telling to provide material for the creation of the Kōka ninja myth, although in a modern study of them Fujita Kazutoshi warns that not all the details are reliable, because in one section shinobi are described raiding Hikone Castle a few years before the place was actually built![24] Section One is of particular interest and considerably more reliable:

The first opportunity for the Kōka Koshi to serve Gongen-sama[25] and provide honourable public service was when Gongen-sama was living in Mikawa Province, and at the time campaigning against Udono Tōtarō 鵜殿藤太郎, an enemy who lived in the same province. In the 2nd month of 1562, using Toda Saburō-dono and Makino Denzō-dono as envoys, he called upon the twenty-one families of Kōka. We soon received the honourable request. Thereupon 200 Kōka-mono 甲賀者 set off for Mikawa Province, and on the 26th day of the same month commenced a night attack on Udono's castle. It resulted in the taking of Udono's head, and his two children were captured alive. Others named among the 200 retainers attacked using fire.

After this they captured the Mikawa Ikkō-ikki bases as far away as the temples at Toro and Harisaki. Gongen-sama was filled with great joy and summoned into his presence the Kōka-mono who had taken part in the attack and bestowed *sake* cups upon them with these words, 'From now on the men of the twenty-one families of Kōka will be treated as intimates; these fellows will not be treated carelessly by the Tokugawa family'. After this they served as expert spies (*onmittsū* 御密通).[26]

The background to the Udono raid was that Imagawa Ujizane had taken members of Ieyasu's family hostage in Sumpu Castle. Ieyasu's plan was to effect the rapid capture of the Imagawa's own Kaminogō Castle and take hostages from the Udono who could then be exchanged for his own family. This section in the petition begs the question as to what the Kōka-shū were doing fighting in Mikawa in 1562 if they were not acting as mercenaries, a role that is also suggested in the elaborate rewriting of the Udono incident in *Mikawa Gofudo ki* of 1833:

> The *monogashira* Mitsuhara Sanzaemon said, 'As this castle is built upon a formidable precipice we will be condemning many of our allies to suffer great losses. Fortunately there are among the *hatamoto* some men who are related to the Kōka-shū of Omi Province. Engage the men of Kōka through this connection and they can enter the castle secretly'. Tadatsugu agreed and beginning with Ban Tarōzaemon Sukeie from Kōka eighty men skilled as *shinobi* 忍に馴たる were ordered to hide in various places, and on the 15th day of the 3rd month they entered the castle in secret, and were soon setting fire to the towers inside the castle. The attackers deliberately did not speak as they ran around killing so the garrison thought they were traitors.[27]

Thanks to the men from Kōka Ieyasu acquired the hostages he needed in the form of the two sons. Following a deal with Imagawa Ujizane the castle was handed back and the sons appointed to its command, but the pair began an unwise policy of supporting the raids into Mikawa Province by the Ikkō-ikki. Ieyasu again entered the field against Kaminogō and once more the Kōka-shū attacked and 'created a disturbance inside the castle'.[28]

Regardless of what precise relationship the Kōka warriors had with Ieyasu, the petition of 1667 illustrates that they claimed to have a long and honourable reputation in the field of undercover operations. This shinobi expertise is mentioned again in Section 3 of the petition. 'At the battle of Komaki-Nagakute in Lord Hideyoshi's reign, Gongen-sama ordered secret operations through Sakakibara Yasumasa as envoy', although we are not told what resulted from it. The section concludes with the disastrous Kii operation when the Kōka Koshi faced 'false charges' and their fiefs were confiscated for generations to come.[29] Section 4 deals with their loyal service unto death at Fushimi and how Natsuka Masaie took hostages, and in Section 5 it is noted that at the time of the siege of Osaka, Yamaoka Dōami supplied ten horsemen and a hundred musketeers from the men of Kōka.[30]

The Shimabara legends
The final section in Akutagawa's 'humble petition' to the Shogun concerns the year 1634, so there is no mention of service by men from Kōka during the largely Christian Shimabara Rebellion of 1637 to 1638, which finished with the siege of the castle of Hara.[31] In Yamaguchi's book *Ninja no Seikatsu* great importance is accorded to the accounts of shinobi operations against Hara, and the mention of Kōka in connection with them is used by Yamaguchi as vital proof for the existence of ninja. Shinobi were certainly present at Hara, although the most reliable accounts of undercover work during the siege have nothing to do with Kōka and instead concern Hosokawa Tadatoshi (1586–1641), the *daimyō* of Kumamoto.

The Hosokawa records present a dispassionate and disinterested account of secret warfare by troops referred to as shinobi no mono and by others where shinobi is used an adverb, but it is interesting to note how narrow the definition of a shinobi skill is in the Hosokawa material. Digging a hole under the wall to provide entry to the castle is not a shinobi operation, even though it provides exciting reading with the defenders thrusting spears down into the gap to flush out infiltrators.

84

The word shinobi is applied only to entering a castle in secret by climbing its walls, not tunnelling beneath them.[32]

In the Hosokawa archives a night raid of the 27th day of the 1st month is carried out by a kogashira ('lieutenant') in the company of Hirano Jibuzaemon. He infiltrates (*shinobi iri*) the castle by night and steals a flag, for which exploit he is awarded five silver coins. The following night another individual, who is described as a shinobi no mono in the company of Yoshida Suke'emon, returns from a raid with rebel heads, but finding that he is being pursued by fourteen or fifteen men he discards all but one of his trophies.[33] On the 13th day of the 2nd month two shinobi no mono sneak into the castle in a very unusual manner because they have ropes attached to them so that if they are shot they can be pulled back, but they return successfully with information. The account then tells us that the Hosokawa had in total ten shinobi no mono within their ranks and that they were used night after night. One is named as the son of the *karō* (senior retainer) Sado. Another is Uehara Heinosuke, who hears the striking of a bell as he enters the castle. Realising that it means that the guard is being changed, he conceals himself.[34] The Hosokawa material therefore reinforces the model of the historical shinobi as élite samurai with particular skills. Their social status must be quite high, because one at least possesses a surname and another is the son of a senior retainer.

In 1960 these events were retold by the modern historian Okada Akio in his book *Amakusa Tokisada* as being performed by 'practitioners of the so-called ninjutsu', and Okada's account continues with two shinobi exploits that are not in the Hosokawa records. First he adds a story about a shinobi no mono falling into a hole in the darkness, then continues with an incident that brings in the name of Kōka for the first time. It relates that when Matsudaira Nobutsuna (1596–1662) went along the Tōkaidō from Edo with reinforcements he stopped at the Minakuchi Post Station in Kōka-gun, where his force was augmented by ten volunteers, a move that Yamaguchi interprets as meaning that he recruited ninja. One of them infiltrated Hara Castle, but was unable to glean much information because he was unfamiliar with the local dialect and the use of Christian expressions. His presence aroused suspicions, but the man escaped in a hail of stones, taking a Christian flag with him as a trophy.[35] Fujita Kazutoshi identifies the source of the Kōka deployment story as a document that dates back no further than 1721 and is repeated in *Tokugawa Jikki* of 1809 in which, in addition to the event described by Okada, there is another raid leading to the capture

NINJA UNMASKING THE MYTH

of thirteen bags of provisions. A further raid is then launched under the protection of covering fire and it is then that one shinobi no mono falls into a hole, as in Okada's other account. They pull back amid a bombardment by stones.[36]

To the ninja enthusiast Yamaguchi the Kōka stories indicated that Matsudaira Nobutsuna called in at Minakuchi Castle and recruited ten battle-ready shinobi no mono, but it is difficult to envisage shinobi being retained in peacetime at Minakuchi for possible use as a rapid deployment squad. Instead Nobutsuna took troops from Kōka simply because it was a fief directly held by the Tokugawa family, and as gaining control of Hara Castle was the object of the entire operation, some of the Kōka men may well have carried out shinobi duties. A desire to emulate the supposed traditional deeds of their ancestors, a common obsession among samurai, would be a sufficient explanation for their enthusiasm to act in this way. The accounts remind one of the Kōka petitions, yet they do have a ring of truth about them because the operations are done so badly. Falling into a hole is not the best outcome for a skilled ninja! Fujita therefore believes that there is some truth behind the claim that Kōka men operated in this way at Shimabara, but that their exploits were enhanced in the later accounts used by Yamaguchi by adding material taken from the Hosokawa records. This is certainly how the stories appear in Yamaguchi's 'catch-all' description in *Ninja no Seikatsu*, which at least reinforces one stereotype about ninja: they were good at stealing things!

Chapter 7

The Shogun's Shinobi

When Tokugawa Ieyasu re-established the position of Shogun in 1603 he chose not to move his headquarters to the ancient imperial capital of Kyoto, but to rule Japan from Edo Castle, the place he had entered in triumph in 1590. This would mean that at precisely the same time that certain loyal residents of Kōka were unintentionally creating their own building blocks for the ninja myth by the petitions to the Shogun, others were creating an entirely different tradition of their own inside Edo Castle. Under the names of the Iga-mono and the Kōka-mono these hereditary guardsmen made a vital contribution to the ninja myth, and when Kikuoka referred to 'the Iga shinobi 伊賀忍 of our day' in 1679 in *Iran ki* he had in mind these men around whom he would weave his narrative of their glorious forebears. As the most trusted retainers of the Shogun the Iga-mono and Kōka-mono patrolled the innermost living quarters of the palace, and their guard duties certainly involved an intelligence-gathering role.[1] In 1626, for example, they conducted a thorough investigation into the *daimyō* of Shikoku and Kyushu.[2] All the *daimyō* in Edo would have become very familiar with the Iga-mono and Kōka-mono who kept tabs on them, so if the two names meant nothing in terms of spying in 1590, by about 1680 they meant everything. Also needless to say, whatever skills the Iga-mono may have possessed, the nature of the Tokugawa State made it impossible for them to be exercised other than on behalf of the ruling family, so any notion of mercenary service was out of the question. The fanciful idea expressed in the movies about Iga and Kōka operating as 'secret societies' (and doing so in broad daylight dressed in black!) is one of the least possible scenarios in the whole of the ninja myth.

The upper-class investigative activities of the Iga/Kōka-mono were part of a huge machine of state intelligence that was part of normal life for the entire population of Japan. As E.H. Norman writes in his study of Japan's emergence into the modern world, 'Espionage was organised on so vast a scale that it has left a heritage of anecdote and proverb, eloquent testimony of the deep mark which it stamped upon the people's consciences'.[3] The operations also clearly strengthened the identification of Iga and Kōka with anything secret, although the reality of their position could be much less heroic. In 1605 the Iga-mono suffered near disgrace in a bizarre reversal of their trusted position when Hattori Hanzō Masanari, the loyal general who had brought them to Edo in 1590, died of illness at the age of 55. He was succeeded by his son, whose name was also read as Masanari even though the characters were different. Young Masanari, whose marriage to the Shogun's niece had given him ideas considerably above his station, knew nothing of Iga Province and treated the Iga-mono more like personal servants than a band of loyal retainers. When he reduced their rice stipends the Iga-mono revolted and barricaded themselves in the temple of Chōzenji, which still stands in the Yotsuya District of Tokyo. Negotiations took place with the 'strikers' and Hattori Masanari was punished.[4]

Another major weakness in the edifice of the glorious Iga shinobi as envisioned by Kikuoka and others is the fact that the word shinobi is only rarely encountered in authentic accounts of intelligence operations mounted by the Edo Iga-mono on behalf of the Shogun. In Shimizu Noboru's highly detailed study of the Tokugawa intelligence system, the word shinobi no mono does not appear at all in the narrative beyond the reign of the second Shogun Hidetada (r. 1605–1622). Instead the Shogun's spies shared a number of euphemisms, and from early in the reign of the third Shogun Tokugawa Iemitsu (r. 1622–1651) we see the first appearance of the word *onmitsu* 隠密 (translated nowadays as 'detective') to describe their intelligence-gathering role.[5] The word *onmitsu* reflects the new situation of peace rather than war, of investigative detective work to build up a dossier on an individual rather than active spying by sneaking into his castle. The *onmitsu* operated within the overall administrative machinery known as the *bakuhan* 幕藩 system, which was composed of two mutually supportive elements: the central *bakufu* 幕府 or Shogunate, and the local government of the *daimyō* territories or *han* 藩. Two groups of officials with similar names oversaw the vital element within it.[6] The *ōmetsuke* 大目付 worked in five pairs and were charged with investigating the *daimyō*. The *metsuke* 目付 monitored the affairs of the Shogun's direct retainers and government officials. There were as many as twenty-four *metsuke* in post at any given time, and although socially inferior to the *ōmetsuke*, they did not report to them but to the superior *wakadoshiyori*, the 'junior elders' who stood second only to the Shogun's council or *rōjū*. One of their responsibilities was to discover if any *daimyō* had broken the rules in *Buke Shohatto*, the written provision for their conduct. Its clauses included making unauthorised repairs to one's castle, arranging unofficial marriage contracts and a host of other minor misdemeanours, including failing to produce a natural heir. This latter requirement, which caused several problems, was eventually dropped in favour of allowing death-bed adoptions. Land transfer was a common punishment for transgressions, and under the first five Tokugawa Shoguns millions of *koku*'s worth of land was transferred, with many *daimyō* being eliminated, created, enriched or impoverished.

In terms of overall state security the Tokugawa Shoguns had two enduring obsessions. The first was Christianity, which had been brought to Japan in 1549 and had enjoyed an early success. Since the time of Hideyoshi there had been fears of a 'fifth column' among the Christian converts who, it was believed, might join forces with a Spanish invading fleet from the Philippines. Remote though that possibility may have been, it was sufficient grounds for the suppression and persecution of Christianity, and the Shimabara Rebellion of 1637 appeared to confirm the Tokugawa's worst fears. The surveillance of suspected 'secret Christians' was undertaken at a *han* level and probably included local shinobi in its operations,[7] although there was a very effective religious arm for the control of society in the form of Buddhism. All Japanese households were affiliated to named local Buddhist temples, thereby giving the Buddhist clergy a policing function over the lives of their registered adherents. From 1671 the temples' bureaucratic duties were extended to the registration of births, marriages, adoptions, deaths, change of residence and occupation.[8] In 1643 an even more severe surveillance network was developed through the system of *gonin-gumi* (five-person groups), whereby households were grouped in units of five and were expected to inform on each other, lest one transgress and all be punished.

By about 1630 the suppression of Christianity was only needed for the lower orders of society because any Christian *daimyō* had by then either apostatised or emigrated to the Philippines. The *metsuke* could therefore concentrate on their other great obsession, which was the potential threat posed by the *tozama daimyō*, the 'outer lords', who had either chosen the wrong side at Sekigahara or who had sat on the fence. The reluctant Tsutsui Sadatsugu of Iga became one such victim when he was dispossessed in 1608. By contrast the loyal *fudai daimyō* received large territories from where they could keep an eye on the possibly disloyal *tozama*. As noted above, some surveillance of the *daimyō* of Shikoku and Kyushu was carried out by the Iga-mono and Kōka-mono in 1626 and their targets were probably *tozama*. The focus was not misplaced, because during the 1850s the *tozama* of Satsuma and Chōshū would become the leaders of the imperial restoration movement. Investigation was also carried out into schools of swordsmanship, and once again the focus was correct because in the years leading up the Meiji Restoration these schools provided cover for young revolutionaries to meet and exchange political ideas while they developed their fighting skills. The Jigen-ryū of the Satsuma-*han* came under particular suspicion and was closely watched.[9]

The government surveillance of the *tozama* was made much easier because of a requirement for the wives and children of all the *daimyō* to live in Edo and for the *daimyō* themselves to alternate their residence between their own castle towns and Edo within successive years, or in some cases alternate half-years. This 'Alternate Attendance System' had two great advantages for the Tokugawa. First, the *daimyō* were dissuaded from rebellion by the presence of the hostages in Edo and second, their expensive processions to and from the capital kept them in a situation of genteel poverty. The requirement that a *daimyō*'s family should live permanently in Edo was in fact quietly abandoned in 1665, but by this time it had become the custom because Edo was the centre of commerce and entertainment and the best place in Japan for what we now call networking. By 1690 five out of every six *daimyō* then living had been born there and some even had to be cajoled by their retainers to make the briefest of visits to the lands of which they were nominally the local lords. The *ōmetsuke* would therefore have had the *tozama* on hand for most of the time, but nevertheless the borders between provinces were controlled by passes and travelling was severely restricted, with the guards on the outskirts of Edo being particularly on the lookout for 'women going out and guns going in' as the contemporary expression had it.

There were also times when a distant *daimyō* needed intelligence of his own, as is illustrated by an amusing anecdote from Kyushu. In 1630 the idea was raised of an invasion of the Philippines as a means of cutting off the supply of clandestine Catholic priests. The instigator of the scheme was the *daimyō* of Shimabara, Matsukura Shigemasa (1574–1630). On 14 December 1630 Shigemasa sent two retainers called Yoshioka Kurōemon and Kimura Gonnojō to Manila to spy out the Spanish defences.[10] They were disguised as merchants and their cover story was that they wished to discuss the development of trade.

While they were away Shigemasa began his military preparations for an invasion of the Philippines. The paucity of sources for what would have been a major international incident is regrettable, although the omissions may simply indicate that certain crucial aspects were never considered. All that is known for certain is that Shigemasa amassed 3,000 bows and muskets for his army.[11] No consideration was given to the need for artillery against the walls of Manila, but that omission could possibly be explained by the fact that Shigemasa was awaiting the arrival of his spies with the relevant information. They returned to Japan in July 1631. No records of the intelligence that they brought back with them have survived, but their information is unlikely to have been either profound or accurate, because they were a far cry from being professional shinobi. The authorities in the Philippines knew exactly who they were and the real purpose of their visit, as is confirmed by the unsigned *Events in Filipinas* of 2 July 1632:

> ...they sent last year in January two merchant ships, under cloak of trade and traffic. Although in Manila warning of this double object had been received, this was not made known; and they were received and regaled as ambassadors from the Tono (lord) of Arima and Bungo. A ceremonious reception and very handsome present were given to them; but the city was put in readiness for whatever might happen.[12]

A separate Jesuit source suggests that a deliberate attempt was made to impress upon the spies the futility of attempting to take Manila by force. It comes in a report sent to Spain on 29 July 1631 by Hernando Perez. In it he states unambiguously that Yoshioka and Kimura were 'sham envoys sent to investigate our situation in order to have an easy conquest of our country'. Perez confirms that presents were given and banquets were held. 'However, although on the surface there was a warm reception, in reality there was a display of military strength in accordance with a situation of war. As the envoys passed through the town the army units were lined up from the seashore to the governor's residence'. Perez concludes that the envoys were 'amazed' by what they saw.[13] Their undercover mission therefore came to nothing, and was nullified anyway by the unexpected death of the invasion commander Matsukura Shigemasa.[14] He had died suddenly in a bath house in Obama while his spies were still in Manila. Murder was suspected.[15]

This amateurish episode contrasts markedly with the professional work done back in Edo when government spies exposed the Rōnin Conspiracy of 1651, a plan to overthrow the Tokugawa government by force. The ringleaders were Yui Shōsetsu and Marubashi Chūya, who were employed as instructors in a sword-fighting school. Their ambitious plan was to choose a windy night and set off an explosion in a government magazine in Edo, thereby starting a conflagration that would spread rapidly through the tightly packed wooden buildings. Under the cover of the fire Edo Castle would be raided and high officials murdered as the start of a major uprising. The plot was betrayed when Chūya caught a fever and started babbling the details of their plans in his sleep. The fact that this led to their arrest strongly suggests the presence of secret agents of some sort, and it is known that the Tokugawa had been monitoring the group for some time. Chūya was arrested in Edo and crucified, while Yui Shōsetsu was apprehended in Sunpu and committed ritual suicide.[16]

Away from Edo all the *daimyō* were responsible for the maintenance of similar good order within their own domains. This naturally involved local surveillance, and the regulations under which a *daimyō* operated required him to have on his staff men who could do this effectively.[17] Matsukura Shigemasa's team were clearly an exception! Quite often these 'secret agents' were called shinobi. The Matsumoto-*han*, for example, had three shinobi in 1726 on a stipend of 18 *koku* each.[18] Other names of local origin were also applied and we do find some interesting examples where shinobi were called Iga-mono or more rarely Kōka-mono. The name is unlikely to indicate that the individuals originated from Iga or Kōka and probably shows that Iga-mono had become a general term for an intelligence gatherer, much as any vacuum cleaner tends to be called a Hoover. This would have arisen because of a widespread awareness of their role in Edo by the *daimyō* in residence, some of whom, of course, were on the receiving end of the Iga-mono's investigations. The *han* of Tsuwano employed men called Iga-mono and there are records of a man called Makino Hachizaemon serving as an Iga-mono in Iwakuni in 1624.[19]

An interesting example of the use of the name shinobi dates from the year 1787 and concerns a man from the province of Tosa called Imakita Sakubei. He had been born in 1742 and succeeded to the family headship when his father died in 1773. Sakubei's ancestor Imakita Jūhei hailed from Iga Province (presumably through the Edo guard unit) and had served under the *daimyō* Yamauchi Kazutoyo as a shinobi at the time of Sekigahara. When the Yamauchi were transferred to Tosa Province Jūhei went with him and the Imakita's descendants continued to act as shinobi on a hereditary basis for many years to come. Sakubei served the Yamauchi in this position on a very modest stipend for thirteen years while holding (and even proclaiming) quite revolutionary political views about the *han*'s structure. Sakubei's persistent idea was that a *daimyō*'s council should include the best people in the domain regardless of their class or status, and that the choice of council members should be made by the people as a whole.[20] He presented these innovative thoughts in the form of a petition to his lord in 1787, and even though it was unsuccessful the document reveals something of the reality of life for a man who bore the title of shinobi during the Tokugawa Period. In his petition he defined the role of a shinobi as being 'to investigate throughout the province the rights and wrongs of all people of high and low status'. As Luke Roberts notes in his article about Sakubei, even though the post was part of the repressive arm of the Tokugawa polity, it had the paradoxical effect of putting someone like him in close touch with all classes of people, a rare occurrence for a samurai in Tokugawa Japan. Sakubei's stipend of 50 *koku* forced him to live in a village, and this close proximity to the common people gave him some sympathy for them, even though he was supposedly acting as a spy.[21]

There were also men called shinobi serving under the *daimyō* of Owari Province, and in 2004 I was introduced to the direct descendant of one of them. Watanabe Toshinobu discovered by chance in 1998 that at least seven generations of his ancestors had served in that capacity from 1679 onwards, beginning with a samurai called Watanabe Hei'emon Toshimitsu. The men were employed by the Owari branch of the Tokugawa in Nagoya along with four other shinobi, and the present Mr Watanabe owns the document that required them not to reveal any secrets about their position. I asked Mr Watanabe how he felt when he first made the discovery. 'To be quite honest', he replied, 'I was a little embarrassed', which is not surprising because nowadays they would probably be called secret policemen and would be regarded with all the fear and loathing that the name usually provokes. The overt nature of their job title may therefore look curious to modern eyes, but it reflects how all-pervading and ubiquitous such low-level espionage was under the Tokugawa regime.

The 'Garden Guards'

In spite of the casual appropriation of their name to indicate any spy, the days of the Iga/Kōka-mono of Edo Castle were sadly numbered, and within twenty years of the start of the eighteenth century their intelligence role had been completely replaced by something far more professional. The development was prompted in 1716 by the death at the age of seven of the Shogun Tokugawa Ietsugu. His brothers were far too young to rule even with a regent, so it was decided to appoint the next Shogun from one of the Tokugawa family's cadet branches. The choice was Tokugawa Yoshimune from the Kii branch of the family, who became the eighth Tokugawa Shogun and reigned until his retirement in 1745. Yoshimune is regarded as one of the best Shoguns, and on taking up the position in Edo he transferred from Kii some of his most trusted officials. These included twenty hand-picked men who specialised in intelligence gathering, to whom he gave the name of the Oniwaban 御庭番, the 'Honourable Garden Guards'. The name probably derives from the location of their barracks in Edo Castle, although the popular view erroneously has it that they were disguised as gardeners in the palace grounds.[22] Edo Castle now had a dedicated and very busy intelligence arm staffed by men who were known personally to the new Shogun. Investigative reports called *fubungaki* 風聞書 were prepared about the *daimyō*, their activities in Edo and in their home provinces. Tours of inspection were undertaken by Oniwaban members in groups of either two or three and these reports would be studied by the *ometsuke*.[23]

The peak of the Tokugawa spies' achievements were two international intelligence operations carried out in 1854 and 1860 respectively in connection with the first official arrivals in Japan for over two centuries. Commodore Matthew Perry of the US Navy had dropped anchor off Uraga in 1853, and the breaking of Japan's self-imposed isolation from the West threw the government into panic. Perry promised to return the following year, and when he did a shinobi from the Tōdō *han* of Anotsu was sent to investigate the ships. As the Anotsu territory included Iga Province, Sawamura Jinzaburō Yasusuke was an Iga-mono in every sense of the word. He carried out his work by being included in an official Japanese party for a reception on board Perry's flagship USS *Susquehanna*. He then returned with information.[24] The descendants of Sawamura still live in the same imposing farmhouse in Iga and take great pride in their authentic shinobi ancestor. Six years later a member of the Oniwaban called Muragaki Norimasa was included in the official delegation to the United States in 1860 and met President Buchanan.[25]

The final record of shinobi in a *daimyō*'s army occurs in 1865. On the fourth day of the fifth lunar month of that year regulations were issued for the army of the Nobeoka *han*, who were expected to join in with the second attempt by the Shogun's forces to crush the rebels of imperialist Chōshū. Article 17 refers to the deployment of shinobi no mono against the enemy. Their duties are 'to pry into the state of the enemy by day and night and on the basis of that intelligence any changes in plans can be considered'.[26] The *daimyō* of Nobeoka therefore probably has the honour of taking shinobi no mono into battle for the last time, but there are no records of how they fared when the disastrous Second Chōshū Campaign began in 1866.

All the above examples remind us that a shinobi was a real person in Tokugawa Japan, not just a character in popular novels. The shinobi out in the provinces and the Oniwaban back in Edo carried out serious duties. The Kōka ancestry of the Watanabe and the Iga ancestry of Imakita and Sawamura also provide strong evidence for a genuine hereditary expertise in intelligence work associated with the two places, an association made even more striking when the terms Iga-mono and Kōka-mono are used in situations where there is apparently no direct historical link with either place. The way in which their undercover operations were carried out was of course very different from the castle-entry techniques of the Sengoku Period, but these men's genealogies strongly indicate that this aspect of the ninja myth, although much exaggerated in the popular view, is based on fact. To some extent the best shinobi really did come from Iga and Kōka.

Chapter 8

Ninjutsu in Black and White

This chapter will shift the focus from ninja to ninjutsu, a topic over which there is a certain sensitivity that is largely absent from discussions of ninja. Ninja enthusiasts, however they may be defined, will usually accept that there have been changes and exaggerations in the ninja's supposedly ancestral accomplishments. They will acknowledge, for example, that the historical shinobi did not dress in black, and when forced to accept the denial of anything else dear to their hearts will take refuge in Kawakami's neat definition of a ninja as a fictionalised shinobi. Problems of acceptance only arise if those same ninja enthusiasts have expended considerable sums of money to be trained in the martial art of ninjutsu and have certificates to that effect, which bear the seal of a grandmaster. Worse still might be the reaction of an initiate who has been entrusted with passing on that tradition as a teacher and has received a copy of the secret scrolls that alone guarantee authenticity.

Faith in a secret process and the beliefs that lie behind it is something that is often encountered in the fields of Japanese art, craft and martial traditions. It is the idea of *hiden* 秘伝, the act of secret transmission within a closed and hereditary group, although the word can also mean the information transmitted in such a manner. *Hiden* knowledge is precious not because it is true in objective terms, but because it has been defined as truthful by successive owners of the knowledge over many centuries. Within a school or tradition its esoteric practitioners convey the essentials of its lore with great care to a few chosen disciples, transmitting not only knowledge, but also power and authority. These intangible possessions are guaranteed by and may also be encapsulated within symbolic objects like scrolls.[1]

To modern recipients of the *hiden* of ninjutsu the historical shinobi was not merely a spy, an infiltrator or a mundane government detective, but a deeply spiritual being whose activities went far beyond the techniques of his trade. Those supposed techniques are nonetheless embraced and practised along with more conventional martial arts by earnest twenty-first century individuals dressed suitably in black, whose vital spiritual development grows in tandem with their physical training. It should not, however, surprise us that a spiritual side should be sought. Since the time of their development during the Tokugawa Period the martial arts of Japan have always had a considerable spiritual perspective compared to similar activities in the West, although the mystical element envisaged for the martial art of ninjutsu goes far beyond any found within judo or karate. It may be expressed in several ways, one of which embraces Japanese religion in a careless manner, as summed up by the front cover of the 2002 book *Iga-Kōka shinobi no subete*.[2] Surrounded by title material and with a small overlay of a modern ninja re-enactor, is a simple stone shrine to Yama no kami, the deity of the mountains.[3] Yama no kami shrines may be found all over Japan, yet the one thing they all have in common is that there is no connection with ninjutsu. The cover picture can only have been chosen to lend an air of mystery to the ninja idea.

Other mystical elements in ninjutsu include the well-known use of esoteric hand gestures, the making of a link with *yamabushi* (mountain ascetics) and even, in some recorded cases, the overcoming of self-inflicted pain. As the editor of the martial arts magazine *Black Belt* once put it, in a profound understatement: 'Ninjitsu [sic] is one of the weirdest martial arts ever invented'.[4] Weird or not, all these approaches derive ultimately from an individual's understanding of what the word ninjutsu actually means, and in modern Iga City the tourist board tries to cover all its bases. In their pamphlet *The Roots of Ninja* ninjutsu is vaguely defined as something that is at once historical, military, mystical and scientific:

> The art of ninjutsu places its emphasis not on force of arms, but upon stealth and using intellectual solutions to combat. One area of ninjutsu called upon divination, psychology and parapsychology to manipulate the enemy's perception, while others called upon disciplines such as astrology and medical horticulture…

This chapter will begin to trace the idea of ninjutsu from the early Tokugawa Period to the twenty-first century. All the points mentioned in the Iga City pamphlet will be encountered as we follow the art of the ninja on its tortuous journey.

Ninjutsu as the lore of the shinobi

The simplest way of looking at ninjutsu is to regard it merely as what Japanese spies did. This is the straightforward approach found in the literature of the Tokugawa Period that dealt with espionage, where ninjutsu simply meant the techniques and the underlying morality of information-gathering contained therein. In other words, 'shinobi no jutsu' (the most appropriate reading of 'ninjutsu' in this case) could either mean 'the techniques used by shinobi' or simply 'undercover techniques'. A very different contemporary interpretation of the word ninjutsu meant magic or sorcery, but the two definitions somehow existed happily side by side within the serious and popular extremes of Tokugawa literature. The magical approach will be addressed in the chapter which follows. The practical side dealt with here was covered by the writers of the nostalgic military manuals of the Tokugawa Period, who shared with their contemporaries who wrote war chronicles a delight in exaggeration and invention.

The results of their endeavours were a number of works that would be fundamental in creating the ninja myth. Some, like *Buyo Benryaku* (1684) and *Buke Myōmokusho* (1806) are generalised military manuals that contain sections on espionage and undercover warfare. They may be dealt with briefly here. *Buyo Benryaku*, where 忍者 is prominently glossed as shinobi no mono, claims a speciality for Iga and Kōka, but consists largely of an extended quotation from Sun Zi defining the five types of spies.[5] *Buke Myōmokushō* 武家名目抄 adds a few new words for shinobi:

> *Shinobi-monomi* 忍び物見 (secret scouts) were people used in secret ways, and their duties were to go into the mountains disguised as firewood gatherers to discover and acquire information about enemy territory. They set off on foot as scouts, many being of *ashigaru* rank, and they were particularly expert at travelling in disguise. Every family used these *kamari monomi* 奸物見 ('wicked' scouts), who were also called *shimi* 芝見 and *kusa*....

Shimi has the same implications as kusa, that of lying low in the surrounding grassland. The passage continues with a unique reference to shinobi acting as assassins:

Shinobi no mono were spies (*kanchō*), also called *kanja* 間者 and *chōsha* 諜者. They travelled in disguise to other territories to assess the enemy situation, they would inveigle their way into the midst of the enemy to discover any weaknesses, and enter enemy castles to set them on fire, furthermore, they also became assassins (*shikaku* 刺) and killed people... All samurai were associated with *ashigaru* of this type who infiltrated in this manner.[6]

Buke Myōmokushō continues with a discussion of *suppa* and *rappa*,[7] so both it and *Buyo Benryaku* give a good idea of how the wartime role of the shinobi no mono was understood in the latter half of the eighteenth century. Yet in neither work do we find any details of the techniques of shinobi no jutsu. For those we have to turn to the volumes that are given over entirely to the topic of espionage in its broadest sense. These fascinating productions, some of which are very well known, are usually regarded as ninjutsu manuals, a term which has a lot of truth in it. During the twentieth century the concepts and images contained in these disparate efforts were extensively mined to feed the ninja myth, and the name once given to the repressive and petty spies of the Tokugawa regime acquired a new flexibility to describe any secret operative from a skilled infiltrator to a cunning thief.

One thing that all the Tokugawa Period shinobi no jutsu manuals have in common is the inclusion of a strong spiritual and religious element, which stresses the correct mindset and the need for loyalty. There are powerful links to Confucianism in this material, and as Confucianism was the underlying philosophy of the Tokugawa Shoguns, it had the additional result of making the books more acceptable to the country's rulers. There is also a common tendency in the genre to attribute the authorship of the work to a hero from antiquity in order to give the book credibility. All the works, however, place their greatest emphasis on successful means of espionage and intelligence-gathering, for which several weird and wonderful techniques and pieces of equipment are recommended. These are the sections of the books that come closest to what the modern imagination regards as ninjutsu, although usually the description tails off with the tantalising statement that 'further information may only be passed on by word of mouth'. These sections were the essence of practical shinobi no jutsu as far as the authors were concerned, who clearly understood the word to mean the techniques of secrecy employed by the people they were describing, coupled with the moral principles that lay behind them. Kikuoka Jōgen shared this straightforward understanding and in a section entitled 'Concerning Iga Ninjutsu' in his *Iga Kyūkō* of 1699 he states how 'the *jizamurai* of Iga were masters of *ninjutsu heihō* 忍術兵法 (ninjutsu strategy)', a concept that would be perfectly realised in words and pictures within these fascinating works.[8]

The practical side of the manuals includes interesting information about herb-based pharmacology and gunpowder technology as it was understood in seventeenth century Japan. Such material alone is worthy of study and is currently the subject of scientific investigation as part of Mie University's collaborative field project. Matters such as the inclusion of Chinese source material, the illustrations of equipment and their revelations about the contemporary use of divination are also valuable in themselves, but have been given added significance because of the use that has been made of them by future generations. Many of the tools and weapons that are illustrated by drawings have been presented to later readers as authentic ninja equipment. The result has been that from the early twentieth century onwards these Tokugawa Period ninjutsu manuals have provided much of the material for every popular ninja book, whether or not its author was fully aware of the fact.

Until quite recently the manuals were widely regarded as secret documents that could only be revealed and passed on to the initiated in the form of tightly wrapped scrolls, a fiction still maintained in several ninja museums where copies of the originals are displayed. One example is the scroll of *Ninjutsu Ōgiden* of 1806 on show in the museum section of the Kōka Ninja House, where a few centimetres of its text lie tantalisingly unrolled. (I purchased my own copy in the souvenir shop of the Chibikko Children's Ninja Theme Park in Nagano Prefecture.) For many years the works were unavailable in modern Japanese and in 1917 Itō Gingetsu believed that his copy of *Shōninki* was the only one in existence, although all have now been published in some form including critical editions with notes and commentaries.[9] Most have also been translated into English by enthusiasts.[10] Sadly, these welcome efforts have often been spoiled by over-editing for commercial reasons. In an attempt to make them more accessible to the ninja fan whole sections have been omitted and what should have been footnotes have been integrated into the texts. Much more seriously, the word ninja is regularly inserted where it never appears in the original. Nevertheless, these translations have been a labour of love and have at least ensured that a much wider audience can learn very quickly that most of their contents are banal, tedious and highly repetitive. The serious student can then home in on what really matters within the Japanese original, and there is much of genuine interest and importance contained therein.

The reader must, however, bear in mind that the manuals' authors defined ninjutsu differently from how it is perceived today. This is most clearly illustrated by a reflection on what the manuals do not contain. The characters 忍者 regularly appear and must be read as shinobi no mono, even though ninja is often forced in by the translator, but more importantly in none of the manuals is there any illustration of the secret operative himself. No masked man in black graces the pages of any of these so-called 'ninja textbooks' either in word or picture, demonstrating that this key element of the ninja myth was a retrospective development. Also absent from the manuals are any references to assassination techniques. The knife in the dark, so beloved of movie producers, bears no relation to the consistent understanding within the manuals of the role of the undercover warrior, which is always intelligence-gathering and disruption as set out centuries earlier in Sun Zi's *Art of War*. There are also no descriptions of the secret operative's personal armoury apart from the sword. *Shūriken*, the spinning 'ninja stars', are totally absent. Similarly, there no references to any one-to-one combat techniques of grappling and sword-fighting as may be found in abundance in contemporary manuals dealing with *bujutsu* and *jūjutsu*. In other words, there are no suggestions anywhere in the manuals that shinobi no jutsu has anything to do with fighting, let alone being a martial art in its own right. That development came about during the myth-making of the 1960s when ninja had to fight each other in movies and something that looked realistic had to be invented quickly.

The reader will also find little mention of any historical undercover operations by shinobi. *Taihei ki* is quoted sporadically, but otherwise nearly all the justificatory precedents for strategy and tactics come from the ancient Chinese military classics, as does most of the theoretical background, which leans heavily on Sun Zi's *Art of War*. Sun Zi's notion of five types of spy often provides the introduction to the topic of espionage in a book's first chapter, which shows how important the *Art of War* was in the formation of the ninja myth.

The early espionage manuals
The earliest of the military manuals containing espionage material is probably the sixteenth century *Kin'etsushū* 訓閲集, which is associated

with the great swordsman Kamiizumi Nobutsuna. As he was a contemporary figure this is more believable than most of the other attributions of authorship. Its third volume deals with spying in a very matter-of-fact way, saying that two or three years before an army is ready to set out on campaign, *kanchō* should be sent on ahead. Craftsmen or merchants with these particular talents should go into the enemy province and assess everything about it, from its topography to the morale of its inhabitants, a matter that can be deduced by how happy or sad the songs that they sing are. The spy is also recommended to make monetary donations to shrines and temples to win the confidence of local priests. Important information about provisions, water supply and geographical features is to be brought back. The spy's 'code of conduct' reflects the admonitions in *The Art of War* in that he should rely only on his own observations, not those obtained at second hand, and that if he is challenged the spy's role is not to fight back, but to return alive with information.[11]

Gunpo Jiyoshū 軍法侍用集, which was begun in 1618 and finally published in 1653, is much more detailed than *Kin'etsushū* and provided much of the material for what came after it, including the famous *Mansenshūkai*.[12] It is supposed to have been based on the strategic writings of the first Shogun Minamoto Yoritomo (1147–99), a fanciful claim considering that the great statesman was not very good at strategy. He certainly made use of spies, but his defeat at Ishibashiyama in 1180 was a humiliation, and from that time onwards Yoritomo left the conduct of military campaigns to his talented brothers.

In *Gunpo Jiyoshū* we see the emergence of a fairly standard formula for the genre of the ninjutsu manual, which is to structure the text around a pastiche of the Chinese military classics along with fairly common-sense advice about spying and descriptions of some clever equipment. Any awkward problems of interpretation are be dealt with by the simple addition of the note that any more information is only to be transmitted by word of mouth. *Gunpo Jiyoshū*'s sixth, seventh and eighth chapters deal with secret operations. The first begins with the heading, 'Why there must be Iga Kōka-mono within every household', and the reason given is, 'if a *daimyō* does not have *settō* (thieves, but glossed as shinobi) serving under him then no matter how good he is he will know nothing of his enemy's dispositions'. This is one of the clearest associations ever made between secret operations and the names given to the men guarding Edo Castle. Like Kikuoka Jōgen, the author of *Gunpo Jiyoshū* explains that in ancient times the men of Iga and Kōka specialised in these skills and passed them on to their descendants.[13] Section Two, which deals with how shinobi (now written as 忍び) should be deployed, says that in ancient times they went in disguise as *yamabushi*, street entertainers, monkey tamers and the like. There is no mention of them dressing in black. Section Three stresses that among spies intelligence is to be valued above physical strength, so there are no superhuman feats there either. There is then a section about torches and explosives and another about night attacks that includes information about scattering caltrops, how to gag a prisoner and the use of ladders.[14] This is all 'good ninja stuff', but *Gunpo Jiyoshū* then moves on to large-scale siege equipment with illustrations of wheeled shield carts made from bundles of bamboo and a huge mobile observation tower.[15] A clumsy wheeled tower could not be further from the image of the ninja, so its inclusion probably indicates that *Gunpo Jiyoshū* is concerned with military intelligence in its broadest sense as the observation of the enemy's positions by every conceivable means.

In complete contrast to the very visible lookout tower is the remarkable inclusion in *Gunpo Jiyoshū* of a curious collection of poems on the theme of the use of shinobi. The poems are known as the *Yoshimori Hyakushu* and are attributed optimistically to Ise Saburō Yoshimori, the loyal follower of Minamoto Yoshitsune from the twelfth century. When translated into English they come out as dreadfully banal because most of their themes are very practical, as in Poem 13 where readers are advised to send out secret scouts to discover the enemy's tactics before devising their own. Other poems advise prayers before beginning an operation, and many of the messages are directed not at the shinobi themselves (for whom the characters for thief are sometimes used), but at the samurai who deploy them, urging them to have no moral qualms about telling lies and using deception in the service of their lords because the ends justify the means. There are also at least two mentions of the requirement that the spies' duties were to convey messages back to their masters so they must not stay and fight, but escape alive. All in all, the poems in *Yoshimori Hyakushu* seem to be a delicate way of getting round the perennial problem of the conflict between the unseemly nature of secret operations and the noble samurai ideal.[16]

The 'Ninja Bible'
Much of the material in *Gunpo Jiyoshū* would be incorporated into the vast compendium known as *Mansenshūkai* 萬川集海, which appeared

in 1676.[17] *Mansenshūkai*, as its title 'ten thousand rivers flow into the sea' implies, is a compilation of different works that brings together various writings on the subject from Japanese sources together with material from the Chinese military classics.[18] It holds a central position in the development of the ninja myth, partly because of its comprehensive nature, but above all for the enormous influence it has had on future generations. After being ignored for centuries *Mansenshūkai* was 'discovered' early in the twentieth century, since which time almost every author writing about ninja or ninjutsu has quoted extensively from it as if it were some repository of mystically revealed truth. It is therefore not surprising that its title appears on a display card in the Iga-ryū Ninja Museum under the appellation of the 'ninja bible'.[19]

The compilation of *Mansenshūkai* is credited to a man from Iga called Fujibayashi Yasutake. The Fujibayashi were known to have been active during the Sengoku Period, and at the end of his introduction the author explains why the book was produced. It was simply 'to record the secrets of these military matters', and as for the motivation that lies behind doing it, he felt the need to preserve the lore before it all disappeared.[20] Yet one also senses a desire to pass on this learning for more practical reasons. The Tokugawa regime relied heavily on surveillance and intelligence gathering, so the compilation of *Mansenshūkai* can be seen as an act of patriotism akin to the enthusiastic demolition of beautiful old castles, which was carried out with such rapidity after the Meiji Restoration in 1868. By destroying these throwbacks to the past the local people demonstrated their rejection of the regime that the new Meiji Emperor had supplanted and showed their loyalty to him by removing any possible future sites of rebellion. The compiler of *Mansenshūkai* is displaying a similar commitment to the political aims of his contemporary rulers (who controlled the nation's entire intelligence network) by putting into the open anything that could possibly be seen as subversive. He may even be trying to ingratiate himself with his betters on the basis of the useful secret knowledge that he crams into the work. *Mansenshūkai* was in fact used as a political tool in a different way in 1789 in a final attempt by the representatives of the dispossessed Kōka Koshi to regain their status as samurai following the failed petitions. A meeting was arranged with a senior official and a bound copy of *Mansenshūkai* was presented to him. It was accepted graciously and the petitioners were given a token gift of silver to take back with them, although nothing was done about any change in status.

Mansenshūkai consists of six major sections, of which the first is entitled *Seishin* 正心 (righteous heart) and places espionage within a moral framework. The essence of the teaching is to have a righteous heart because the secret techniques involve conspiracy and deception. If one cannot prevent one's own heart and mind from being deceived, one cannot carry out one's mission. A righteous heart is acquired through the pursuit of justice, and Confucius is quoted so that the emphasis may be laid on the loyalty that exists within a Confucian hierarchy. Ninjutsu must not be used for one's personal gain, so a strict distinction is made between the noble calling of the shinobi and the sordid world of the thief. A distinction is also made between the shinobi's activities and the way of the samurai. Just as in *Gunpo Jiyoshū* the shinobi (who is simply understood as the 'living agent': Sun Zi's fifth type of spy) must extricate himself alive from a situation in order to take back the intelligence for which he was sent on his mission in the first place. The shinobi's world is therefore a topsy-turvy one. Deception and fake loyalty are honourable acts when they are applied to an enemy, and unlike the noble samurai who glories in his own achievements the shinobi must maintain his anonymity, even to the extent that his own comrades must never know what his real function is. There is no compromise on the traditional need for loyalty and the willingness to risk death; the distinction lies merely in the way these vital principles are applied.

The second section is entitled *Shōchi* 将知 (military intelligence), and includes among other things some practical recommendations on how to deal with Sun Zi's double agents. Chapters follow about the two practical ways of spying called *yōnin* 陰忍 and *innin* 陽忍. *Yōnin* is the situation where the spy's physical presence is known to his victims because he is either in disguise or has been accepted as part of a community. The techniques of *innin* are those of invisibility, involving infiltration under cover of darkness and the like. The distinction between *yōnin* and *innin* is illustrated by an anecdote about one of Iga's legendary shinobi who entered Sawayama Castle in the disguise of a woodcutter, but then had to hide under the floorboards. In other words, he moved from *yōnin* to *innin* within the same operation.

The fifth section is *Tenji* 天時 ('the times of heaven'), which is an interesting blend of astronomy, astrology and divination, while the final section is entitled *Ninki*, 忍器 (secret weapons). These, and the pictures of them, have made *Mansenshūkai* famous, and just as in *Gunpo Jiyoshū* there is page after page of equipment for information-gathering operations. They include saws, skeleton keys and flotation devices, all of are repeated endlessly in twentieth-century books about ninjutsu. The most notorious is the *mizugumo* 水蜘蛛, the wooden 'water spider' that is commonly believed to allow a ninja to walk on water. It has long been regarded as a pair of huge water shoes, and the Iga-ryū Ninja Museum once showed its use in that way. In fact it is a simple flotation device on which one sits. Far more authentic are the sections in *Mansenshūkai* and other manuals about medicines. Kōka was an early centre for pharmaceuticals, which suggests an origin for the claims about these aspects of ninjutsu being based on science.

Mansenshūkai is a large work and much of it makes dull reading unless it is enlivened by imaginative commentators desperate to find within it the factors that modern movies have led them to regard as authentic ninjutsu. Such careless retrospective interpretations have produced exaggerations in the past, for example the widely accepted notion of the existence of a ninja hierarchy of upper (*jōnin* 上忍), middle (*chūnin* 中忍) and lower (*genin* 下忍) ranks. These expressions are included in *Mansenshūkai* to indicate differences in skill levels, but in 1956 the imaginative author Okuse Heishichirō transformed the idea into a ranking system. This notion has since become widely accepted and *Mansenshūkai* is cited as proof of the concept.

Even more contrived is the use made of a brief reference in *Mansenshūkai* to 'the enigmatic phrase' 久ノ一術 *kunoichi jutsu*, which is commonly taken to mean the activities of female ninja, even though there is no suggestion of this in the original. It is to the 1950s novelist Yamada Futarō that we owe the elaboration of the meaning of *kunoichi* from 'female' to 'female ninja'. It is probably to the 1950s novelist that we owe the elaboration, including the idea that the character ku is a homophone for *kyū* 九 (the number 9). *Kunoichi* (nine plus one) is therefore a vulgar phrase for the word 'female', because a woman has ten orifices in her body compared to a man. Alternatively it is a word produced by cleverly dividing up the character 女 into the three elements of *ku*, く *no* ノ, and *ichi* 一.[21] The existence of female ninja is now so widely accepted that it is unthinkable to have a modern ninjutsu demonstration without the presence of a *kunoichi* to add her own version of ninja girl power.

Mansenshūkai includes only one historic Japanese example of undercover work and there are many errors in its telling. It is a strange incident concerning an operation against a man who bore the splendid family name of Dodo. *Mansenshūkai* tells us that Dodo was a treacherous retainer of the Rokkaku of Ōmi Province. He was entrusted with the castle of Sawayama but took it over for himself, so in 1559 Rokkaku Jōtei hired the services of a leader of shinobi called Tateoka Dōjun, who used Iga and Kōka shinobi in disguise (a technique called *bakemono jutsu* 妖者術) to infiltrate the castle where Dodo was killed.[22] Because of its inclusion in both *Mansenshūkai* and the other manual *Shinobi no hiden*, this has become one of the best known of all 'ninja operations', but most of the story is unhistorical, including the date.

In reality Dodo Oki-no-Kami Kuranosuke was a retainer of the Azai and not their enemies the Rokkaku. He had been entrusted by Azai Nagamasa to hold Sawayama Castle when Rokkaku Jōtei invaded northern Ōmi in 1559. Sawayama was a rough *yamashiro*, as shown by the presence at its foot of the Dodo *yakata* (mansion) where Dodo would normally have lived.[23] Rather than using shinobi to attack him Rokkaku Jōtei attempted to bribe Dodo by offering him a vastly increased fief and personal ownership of Sawayama Castle on becoming a Rokkaku vassal, but Dodo nobly refused.[24] In 1560 Dodo fought once more for the Azai against the Rokkaku at the siege of Hida Castle, a place where their territories met. This was an interesting attack carried out using flooding, and it concluded with the subsequent Battle of Norada. Rokkaku Jōtei attacked Sawayama Castle again in 1561.[25] It was at this point that any shinobi attack would have taken place, but none is recorded in the chronicle *Azai Sandai ki*, and all that is known for certain is that Dodo Oki-no-Kami committed suicide as a result of the raid.[26] The version of the story included by Kikuoka Jōgen in his *Iga Kyūkō* of 1699 gets Dodo's affiliation right, but he also accepts as real the *Mansenshūkai* account of his local heroes the Iga shinobi in action:

> In the middle of the Eiroku era, when Dodo Echizen-no-kami was besieged in Sawayama Castle in Ōmi Province, Dōjun of Tateoka village, became Sasaki [Rokkaku] Jōtei's ally, and 44 masters of ninjutsu from his disciples in Iga and four men from the Kōka-ryū, 48 men in total, secretly entered (忍入) the castle and Dodo was killed. By this, the name of Iga became famous and stood out as the most distinguished tradition of ninjutsu.[27]

The later military manuals
Mansenshūkai was followed by a number of shorter works, of which two are particularly important because they make up the 'big three' ninjutsu works of the Tokugawa Period and are almost as extensively quoted.

The first is the comparatively short *Shōninki* 正忍記, which was written by Natori Masataka in 1681. He was a retainer of the Kii *han* (modern Wakayama Prefecture) and it is traditionally believed that *Shōninki* provided useful information for the Shogun's Oniwaban when they moved from Kii to Edo. It begins with a short historical introduction and a disavowal that ninjutsu is anything to do with thieving. *Shōninki*'s explanation of the types of spies comes straight from *The Art of War*. One familiar 'ninja trick' mentioned for the first time in *Shōninki* is that of using a sword guard as a footrest to climb a wall. Other material includes judging a person's character by his appearance and a lesson in palm-reading.

Shōninki is famous for two particular sections. The first is a list of the six essential tools of the trade: a straw hat, a grappling hook, a stone pencil, medicines, a towel and a tinder box. It recommends dark colours for one's clothes and the use of a *wakizashi*, the shorter of the two swords usually carried by samurai. No reason is given, but it is probably the first written mention of the usual understanding of the traditional ninja sword as being short. The other much-quoted section concerns disguise and there are seven recommended choices, all of which would allow a spy either to blend in with a crowd, to approach a gathering unobtrusively or, more crucially in Tokugawa Japan, to be allowed through checkpoints. The disguises include three religious figures: the *komusō* (an itinerant flute-playing Zen monk), an ordinary monk and a *yamabushi*, a mountain pilgrim believed to have demonic powers of exorcism. Historically the *yamabushi* were able to cross closely patrolled provincial borders more freely than many other individuals and were almost unique in being allowed to carry swords, so it was almost the ideal cover for a role that involved cross-border espionage. In an urban situation the disguise of a *hōkashi* (street entertainer) or *sarugaku* (actor) would be accepted at face value and allow the spy to gain information from overheard conversations, but *Shōninki* emphasises that disguise alone is not enough. The undercover operative should be thoroughly versed in every role including the vital skill of being able to mimic local dialects.

Shinobi no hiden (also read as *Ninpiden*) 忍秘伝 appeared sometime around 1700, although its authorship is credited to Hattori Hanzō, an attribution which links it to one of the best known names in the ninja myth. Its real author is described as an Iga-mono from the Okayama-han, so we may regard him in a similar light to Imakita Sakubei of

Tosa.[28] *Shinobi no hiden* begins with a short historical introduction and then jumps straight into the now well-known ninjutsu toolkit with accompanying illustrations. There is information about which people to target for information gathering, how to deal with dangerous dogs and the first mention of the technique of walking silently by placing one's feet on one's hands, a process which is always the comic highlight of any modern ninjutsu demonstration.

In 1787 Chikamatsu Shigenori of Owari produced *Yōkan Kajō Denmoku Kugi* 用間加条伝目口儀, which adds little to the above,[29] unlike the last of the genre, the fascinating *Ninjutsu Ōgiden* 忍術応義傳 (the righteous response to ninjutsu), which appeared about 1800. It consists of only one scroll, making it the shortest of the genre of writings on ninjutsu, but it is also the most esoteric in terms of its religious references. There are two versions of *Ninjutsu Ōgiden*, the one translated fully below and a slightly longer version that differs from the shorter one by little more than the addition of a preface. According to Yamada, the versions are associated with two of the 'Fifty-three families of Kōka', the longer being linked to the Tongu and the shorter to the Mochizuki.[30]

Like most of the other ninjutsu manuals, *Ninjutsu Ōgiden*'s authorship is credited to a historical figure. These details are provided at the end with a date of 11m 7d Tenshō 14 (17 December 1586) and the name Mochizuki Shigeie 'a descendant of Kōka Saburō Kaneie'. To credit the authorship to Mochizuki Shigeie is to give the work an unparalleled authenticity, because Kōka (Mochizuki) Saburō was a semi-legendary figure regarded as the forefather of the Mochizuki. He lived during the tenth century and took part in the suppression of Taira Masakado's rebellion. His deified spirit is enshrined at the Aekuni Shrine in Iga.[31]

There is nothing in *Ninjutsu Ōgiden* about secret techniques for entering castles or spying on enemies, and anything remotely connected to such topics is instead covered by the usual reference to oral transmission. The work as a whole is instead an exhortation to virtue written from the point of view of someone who is supposedly immersed in the secret techniques. The tone of the document is that of passing a tradition on to an initiate, and scrolls of *Ninjutsu Ōgiden* have been used in the past as a way of certifying an individual's worth as a successor. Fujita Seikō, one of the greatest names in ninjutsu during the twentieth century, is believed to have been given *Ninjutsu Ōgiden* by his grandfather. It is the epitome of ninjutsu morality and translates as follows:

> An exposition of the righteous response to *ninjutsu*
> It is a faithful saying that from no man should ordinary laws be hidden as the land is within the beauty of daybreak, because whatever befalls there are benefits gained. For this reason I will reveal the hidden evidence of secret beliefs and their origins. A general abides in illustrious virtue, while for a soldier it is to be found in obedience to the highest good, he is resolute in life and death as from ancient times. Peace begins with the army, so seek and achieve righteousness in your life and none will be poor in his heart, no matter how meagre his reward.
>
> Your humble servant has proficiency in this, because I have not parted company with my loyalty and successive acts of righteousness are the way to heaven, nor have I any desire to enhance my military renown in addition to my meritorious deeds, my simple plan is of the highest virtue in that when these techniques are revealed naturally and without concealment, the weakest will overcome the strongest, just as it was, if we reflect upon it, in the case in the battle of the Gods and the Asuras.
>
> Under the command of Marishiten and Kongōrikishi the Asuras were brought to heel by employing secret strategies (秘法), and it is also said that Japanese and Chinese military ways and other mysterious techniques (妙術) are to be found among the military techniques of Crown Prince Taishi.
>
> After the subjugation of Moriya we acquired a name for which the character *shinobi* (忍) is used, and although the ancient teachings were lost in the time of strife, among them twelve or thirteen that were mainly concerned with matters of the quelling of evil spirits and explanations about the moment of entering into Buddhahood, were written down and became the authentic principles of the Kōka School. In the not too distant future those who gain victories by means of gunpowder devices will enquire of you about these techniques, but you must transmit these secret teachings only to a follower who is an honourable individual with this natural capacity and make no exceptions for anyone on the basis of their

rank; apart from such a person you must not teach this priceless treasure.

Shōtoku Taishi naturally prayed for victory to the Great Shrine of Amaterasu, and as he was at the time on the lowest of the three peaks that are within the three provinces of the Bay of Ise, that is to say the mountain top on which Aburahi Daimyōjin once descended from heaven, it is said that ever since then votive lights have been offered, and new ones are presented by the local people even to this day, Taishi prayed during his sojourn in the shrine and performed religious rites, and in *Myōdōjutsu sho*, the history of Aburahi Daimyōjin, it says that the reward bestowed upon him was that an army of the people from both provinces assembled to subjugate his powerful enemy Moriya. So it is in the world of today, the man who dedicates himself to the Treasure Hall of Aburahi Daimyōjin, who understands its mysteries, who by constant entreaty prays to the utmost with his whole heart, who will not put his own life first, who endures patiently, who will not give up, he is the man to whom this mysterious righteousness will be entrusted as if they were the deeds of a general, and swiftly will his boat sail on the water, but again he must be prudent lest his boat capsize.

The Heart
It has ever been that only in the heart are righteousness, sincerity and loyalty in harmony and made marvellous through prayer.

The Body
The entire practice of ninjutsu is contained in seven principles, these are the only ones that you need to be aware of.

Providing for and respect for the army
The *maki* 巻 (book or volume) of secret weapons
The *maki* of night-time operations
The *maki* of the way of secrecy 忍道
The *maki* of fire and water

These are the five texts handed down by word of mouth.

The above five *maki* are not handed down in writing but by word of mouth.

111

(There follows a pictorial diagram illustrating the cardinal points as the azure dragon, vermilion bird, white tiger, black turtle, yellow dragon of earth)

The seventh day of the eleventh month of the fourteenth year of Tenshō
The descendant of Kōka Saburō Kaneie
Mochizuki Shigeie

(Red seal)

The historical and religious references in *Ninjutsu Ōgiden* that underpin its moral tone are considerable. The mention of Crown Prince Shōtoku is to his successful quelling (with divine help) of a rebellion by Mononobe no Moriya in 587. Aburahi Daimyōjin, one of the gods to whom he prayed, was the tutelary deity of the Kōka families, along with being the *kami* of oil and lubricants, and is still worshipped at a shrine in Kōka. Amaterasu is the goddess of the sun who was the founder of the imperial line. Kongōrikishi is a guardian god, and the mention of a heavenly battle refers to the legendary conflict in Hindu mythology between gods and Asuras ('titans' or demi-gods), but the most interesting religious reference is to Marishiten, whose worship tradition lies at the interface between Buddhism and folk religious practices.[32] Marishiten (Mārīcī in India), has acted in many roles over the centuries in both a female and a male form, and she is often shown as multi-armed and multi-headed, standing on a wild boar. The deity offered a unique blessing to a warrior because Marishiten was the god to whom one prayed if one wanted to become invisible. Such a belief may sound surprising among samurai, because the war chronicles suggest that once on the battlefield they made every effort to be seen. Yet the invisibility sought from Marishiten was not necessarily of a physical nature. It could indicate a range of techniques from hiding one's strategic or tactical intentions, to the psychological confusion of an enemy swordsman during hand-to-hand combat. A warrior effectively became invisible if his enemy's mind was confused and his own was clear, and it was to these qualities that a Marishiten devotee aspired. There are also descriptions of Marishiten standing with her back to the sun when she fought the Asuras so that her enemies were temporarily blinded.[33]

It is not difficult to see how the cult of Marishiten could be associated with undercover operations.[34] The invisibility implied by the

112

understanding of the *nin* in ninjutsu as secrecy fits perfectly. In practical terms most Marishiten-orientated rituals were the utterance of relatively simple prayers, but they also included supposedly magical spells and the use of *mudra* (esoteric hand gestures made by twisting the fingers in various patterns). The latter are quite common in Japanese Buddhism, but have received great prominence in popular culture because of their association with ninja. No textbook on ninjutsu is complete without some reference to the *kuji in* 九字印, the mystical nine hand gestures that the secret warrior would make, either to confuse his enemies or to achieve invisibility. The actions and their symbolism are of Chinese origin. Each hand gesture is associated with a Chinese character pronounced in Japanese as *rin, pyo, tō, sha, kai, jin, retsu, zai* and *zen*. At the same time as uttering the sounds the devotee cuts the air with his arm as if with a sword. The ritual concludes with a sign that signifies Buddha's wisdom. The traditional idea was that by performing this ritual one erected an invisible barrier around one's body that would protect against evil spirits and drive them away.[35] Popular books state that they were used by a ninja to hypnotise an enemy into inaction or to give the ninja a sudden surge in power, and having seen them demonstrated to great dramatic effect by Kawakami Jinichi I can confirm that they convey considerable authority at the very least.

In conclusion, even though the shinobi no jutsu manuals of the Tokugawa Period have exerted a huge influence on the ninja as he is presently understood, it is difficult to assess what impact they had on the target audience for whom they were written. One suspects that it was quite small, because their authors were describing the techniques of an age that had apparently passed: skills that would quite obviously be totally unnecessary under the benevolent rule of the Tokugawa Shoguns. In that political sense at least they needed to be presented as curios from a less favoured age.

In contrast to the ninjutsu authors, a handful of daring military writers thought more widely and looked into Japan's strategic position within the isolation which Tokugawa policy had brought about. It became obvious to them that Japan was not only lagging behind in military development, but its exclusion from the outside world also prevented any progress towards parity being made. Sometimes these writings were suppressed and the printing blocks were destroyed, only to be carved again if a Russian ship was spotted in Japanese waters. Experiments then began in the field of military technology and coastal protection, a matter in which the arts of the shinobi were necessarily limited. Intelligence work would only come into its own when contact was eventually made with the foreigners, as Sawamura's shinobi mission to Perry's fleet would indicate, but until that time internal concerns prevailed, and that aspect of Tokugawa governance was covered very efficiently by the formal spy network that allowed for no privacy in the daily lives of Japan's population and no need for anyone to climb into anyone else's castle. The techniques and tools in *Mansenshūkai* were by that time quaint anachronisms, waiting to be discovered during the twentieth century by those seeking an understanding of ninjutsu, when they would be hailed as proof of the essential continuity vested in the ninja myth.

Chapter 9

The Magic of Ninjutsu

The writers of the military manuals may have regarded ninjutsu as the techniques of information gathering, but to the ordinary man in the streets of Edo the word ninjutsu indicated instead various supernatural tricks, like shape-shifting, sorcery and invisibility, and the Tokugawa Period's love of the supernatural produced a written interpretation of it in which magical ninjutsu was performed by fictional characters whose exploits thrilled their readers. The tales first appeared in popular novels and plays and were given additional visual impact through the craze for mass-produced woodblock prints called *ukiyo e* ('pictures of the floating world'). The stories' influence on the ninja tradition began when their subjects reappeared in the pulp fiction and films of the early twentieth century. It was at that stage that ninjutsu magicians blended with shinobi and began their transformation into ninja.

A hero called Tobi ('flying') Katō (Katō Danzō) probably has the honour of being the first completely fictional ninjutsu performer in Japanese literature. He made his debut in Azai Ryōi's *Otogi boku* of 1666. Tobi Katō is supposed to have lived during the sixteenth century and his extraordinary abilities included swallowing a live bull in front of an audience. He is also to be found engaged as a more mundane 'shinobi thief', stealing people's armour.[1] The practice of ninjutsu is even more magical among the later fictional characters whose overall *yōjutsu* (sorcery) was based around shape-shifting using various 'jutsu' bearing the names of toads, butterflies and rats. The character Tenjiku ('India') Tokubei is supposed to have been based on an adventurer from the early Tokugawa Period who wrote an account of his voyages as far as India. The Tokubei of the kabuki theatre had the ability to disappear at will and to transform himself into a giant toad using 'toad techniques' (*gama no jutsu*). Jiraiya, (literally 'young thunder') a hero from popular literature since 1806, was also a shape-shifter said to be able turn himself into a toad or on other occasions to ride one. In the novel *Jiraiya Gōketsu Monogatari* he falls in love with a beautiful girl who practices slug magic and can transform herself into that very unattractive gastropod. His enemy Orochimaru can change himself into a snake.[2]

The sorcerer Nikki Danjō's speciality was to transform himself into a rat, and he does so in the kabuki play *Meiboku Sendai Hagi*, written originally for the puppet theatre in 1777.[3] The plot hinges around the existence of a scroll bearing the names of a gang of conspirators. Nikki Danjō appears in the form of a large grey rat, grabs the scroll in its mouth and runs off with it. A guard sees the rat scuttling past and wounds the rat on the head, but it wriggles away and escapes down a trapdoor in a cloud of smoke. The restored Nikki Danjō then rises from the floor like the demon king with the scroll still clamped between his teeth. Prints of Nikki Danjō's entrance are commonly reproduced in books about ninja with no explanation of the context, so that it may be thought that to hold a scroll in one's mouth was part of magical ninjutsu, while in reality the scroll is there simply because Nikki Danjō has up to that point been a rat.[4]

The contribution that the kabuki theatre made to the ninja myth was largely confined to magicians like these who would be transformed into ninja many centuries later, although kabuki may also have been responsible for one very important element in the ninja's image: his black costume. To wear black is of course in complete contradiction to the advice in *Shōninki* that the shinobi should wear a disguise to blend in with his surroundings or be accepted by his victims, but the black garb has become central to the image of the ninja. The stage hands who moved the scenery and assisted the actors to change costumes in full view of the audience were dressed in black with masked faces as a visual shorthand for invisibility. They were called *kurogo* (or *kuroko*) 黒子. A similar dress code may be noticed in the Bunraku puppet theatre, and of course the black costume shared with many other world cultures the 'cloak and dagger' notion of the concealment of one's identity and the ability to merge into the shadows.

At first sight the lack of mention of black clothes in the ninjutsu manuals appears to be both contradicted and remedied by contemporary woodblock printed books and *ukiyo e* prints, which just happen to be peppered with images of men in black doing things in secret. Modern minds clouded by the movie image of the ninja inevitably see the figures as ninja, and it is very tempting to juxtapose the espionage tools from the manuals with the men in the book illustrations and present the resulting blend as historical proof for the existence of ninja. This was a trap into which I fell in 1991, but the reality of the situation is not so straightforward.

These illustrations, of which examples are discussed below, would provide the missing visual element for the ninja that twentieth century writers were creating who would eventually combine both words and pictures to give us the ninja. The Tokugawa Period illustrations that are commonly interpreted as being of ninja are of two kinds: those showing ninjutsu magicians where there is no visual connection with the ninja image, and those showing secret operations (usually assassinations) where the protagonists are dressed in black to a greater or lesser extent. A very famous and simple example of the latter is a much reproduced illustration in the *Manga* (Sketchbooks) by Hokusai (1760–1849). Hokusai published the *Manga* as a series over a period of several years from 1814 onwards, and the figure appears on a page along with other martial arts activities. He is dressed in black and is climbing a rope, but because it is only a sketch there is no further information provided, although its juxtaposition with the other figures who are performing *jūjutsu* suggests that Hokusai is saying, 'these are the sorts of things that real martial artists do'.

A ninja-like figure also appears in *Nise Murasaki Inaka Genji*, a parody of the *Tale of Genji* published in Edo between 1828 and 1842 with illustrations by Kunisada. Its hero is called Prince Mitsuuji. In one picture the prince appears to be lost in the rapture of his *koto* playing and the brilliance of the moon, oblivious to the fact that a black-clad assassin is behind him with a drawn sword. Yet Mitsuuji is aware of his presence and takes full control of the situation, because in the next picture he pins the man to the ground using a *jūjutsu* hold while taking the assailant's sword triumphantly in his other hand.[5] The man shown attacking Prince Mitsuuji is dressed identically to the image we now have of the ninja. He is however an assassin, not a spy, and his black garb is not intended as a uniform for a shinobi.

Much more intriguing is an enigmatic illustration in *Ehon Taikō ki*, a historical romance based on the life of Hideyoshi and illustrated by Okada Gyokuzan in 1802. In one of the later books in the series is a picture that modern eyes inevitably identify as a ninja both through his appearance and his behaviour. The man's name is given as Kimura Hitachi-no-suke 木村常陸介, a historical figure who was a retainer of Toyotomi Hidetsugu. The childless Toyotomi Hideyoshi had chosen Hidetsugu as his heir, but when Hideyoshi's favourite concubine finally bore him a son Hidetsugu was disinherited and eventually forced to commit suicide. In this illustration his loyal follower Kimura is climbing into Fushimi castle with the intention of assassinating Hideyoshi, and the title of the chapter reads 'The assassination plot of Kimura Hitachi-no-suke'. The story may be fictitious, but Kimura was certainly faithful to the dispossessed Hidetsugu and would follow him to death by committing suicide in 1595.[6] Kimura is shown crossing the moat suspended precariously from a hooked rope with a perfect collection of 'ninja weapons' dangling from his belt. He is dressed in black, is masked and shows only his eyes. The accompanying text claims that 'he had trained in *bujutsu* and was adept in *shinobi* (*shinobi ni myō o etari*)'. He entered secretly (shinobi iri) into Osaka Fushimi Castle 'by means of the ninjutsu he had learned'.[7]

It is a wonderfully convincing picture, yet instead of being proof for the existence of ninja in 1802 it shows instead that the modern image of the ninja is based on pictures like these, because there is no intention of depicting Kimura as a dedicated *shinobi no mono*. He is an opportunistic assassin operating under cover of darkness, an operation for which a black costume would be an excellent choice, and the only concessions towards the modern vocabulary of the ninja are the references to him being skilled in (or as a) shinobi and having learned ninjutsu. It may be a very fine distinction, but just as in the example of the young Hideyoshi at Kaizōji quoted earlier from *Shinchō-Kō ki* the subject is a named individual acting in a skilled and secretive manner, not a specialised infiltrator.

Another assassination attempt by a man in black is the subject of an exciting 1883 *ukiyo e* print by Toyonobu, and once again the man is not on a spying mission and is not anonymous. His name is Manabe Rokurō and he is trying to kill Oda Nobunaga. Oda Nobunaga destroyed the *daimyō* Hatano Hideharu in 1573 and Manabe Rokurō was given the task of revenge. He tried to sneak into Nobunaga's castle of Azuchi to stab Nobunaga while he was asleep in his bedroom, but was discovered and captured by two of the guards. He then committed suicide, and his body was displayed in the local market place to discourage any other would-be killers. To our eyes his dramatic portrayal by Toyonobu is the perfect ninja image, but the black costume is once again an assassin's cover, not the uniform of a shinobi.

So what are we to make of these pictures? They are, after all, illustrations of people who are skilled in shinobi activities and who dress in black to carry out their missions. I believe that they demonstrate that the visual image of the men whom we now call ninja developed in parallel to the written descriptions of them that appeared at about the

Above: The image of the 'benchmark ninja' of the movies is perfectly illustrated by this life-size figure at the Toei Film Studios Park in Kyoto. He is dressed from head to foot in black; his face is masked, his sword is slung around his back and he is about to throw the dart-like version of a *shūriken*.

Below: The ninja are largely associated with the former Iga Province. Iga-Ueno Castle is seen here from the site of Hijiyama Castle, where one of the fiercest battles of the Iga Rebellion took place in 1581. The heroic events surrounding this war would provide many of the building blocks for the future ninja myth, particularly through the writings of Kikuoka Nyogen, whose *Iran ki* of 1679 provided Iga with a great work of epic literature.

This fine bronze statue of a ninja stands outside Kōka Station in Shiga Prefecture. He is carrying a circular pestle for grinding medicines. During the time of civil wars the Kōka area was closely associated with its neighbour Iga. Its *jizamurai* (warriors of modest means) cooperated in the resistance against Oda Nobunaga during the 1560s and 1570s, adding greatly to the ninja myth.

One model for the origins of the ninja lies in the activities of lower class warriors who operated outside the law. They were recruited by *daimyō* to carry out raids and surprise attacks. Various names were given to these men. The Hōjō family called them *rappa* and had a considerable respect for their talents. In this illustration from *Hōjō Godai ki* we see the *rappa* leader Kazama Kotarō supervising a disruptive raid on an enemy camp. He was supposed to be a giant of a man, but this book illustration has been defaced by adding graffiti, making him look demonic.

The Kizugawa as seen from the site of Kasagi Castle. An account of a raid on Kasagi Castle in 1541 tells of warriors from Iga and Kōka carrying out an attack and burning buildings. The passage has often been cited as proof for the existence of ninja and their mercenary activities. However, it is more correctly understood in the context of the independent landowners of Iga asserting their position by involving themselves in neighbouring conflicts. Short-term operations like this gave rise to the notion of a ninja.

The Shogun Ashikaga Yoshihisa (1465-1489) is popularly regarded as the first noble victim of a ninja operation. The incident happened in 1487 and is called the Chōkyō no Ran from the year period. Yoshihisa moved against Rokkaku Takayori who was busily appropriating his neighbours' lands for himself. The Rokkaku territories included Kōka, and lower-class warriors from Kōka raided the Shogun's camp at Magari by night. The nature of the attacks delivered against Yoshihisa led the compiler of *Ōmi Onkoroku* in 1684-88 to describe it as the first operation by 'the Iga-Kōka *shinobi no shū* who are spoken of highly throughout the world'.

Left: The military genius Oda Nobunaga (1534-1582) was the first of Japan's unifiers whose campaigns brought the Sengoku Period to an end. In 1568 the Rokkaku family, assisted by the *jizamurai* of Kōka, frustrated his attempts to move through Ōmi Province and on to Kyoto. A sporadic war ensued and Nobunaga finally received their surrender in 1574. Kōka then passed under Nobunaga's jurisdiction.

Below: In 1576 Oda Nobunaga asserted his authority over Ōmi Province by building the magnificent castle of Azuchi beside Lake Biwa. It was not far from the Rokkaku's former fortress of Kannonji, and, to add insult to injury, Nobunaga dismantled the Rokkaku's family temple of Chōjūji in Kōka and moved it to Azuchi to create the temple of Sōkenji. Most of Azuchi Castle was burned down when Nobunaga died in 1582, but the three-storey pagoda and this gate survive to this day.

Right: In 1579 Nobunaga's son Oda Nobukatsu moved against Iga Province at the start of the campaign that became known as the Iga Rebellion. The Iga *jizamurai* responded with a largely guerrilla campaign. Nobukatsu repaired the fort of Maruyama, which the Iga men attacked from their base on Tendōzan, the hill which is now the site of the temple of Muryōjufukuji, shown here from across the river at Maruyama. The same hill would be used as one of Iga's two concentration points by its defenders during Nobunaga's second invasion in 1581.

Above: Oda Nobunaga sent a second army against Iga in 1581. Some of the fiercest fighting took place here at the castle of Hijiyama. Hijiyama was so vigorously defended that Nobunaga's men had to resort to setting fire to the entire wooden complex. It is now the site of a Buddhist temple.

Above: The final battle of the Iga Rebellion took place at the castle of Kashiwara. Its stubborn defence, which included raiding and the use of trickery, allowed Kikuoka Jōgen to write some of his best descriptions of the irregular warfare that would become the building blocks for the ninja myth. Surprisingly for so fierce a struggle, the site of Kashiwara ended with peace negotiations.

Below: The Aekuni Shrine in Iga City has always been the *ichinomiya* (principle shrine) of the former province. The place was burned to the ground by Oda Nobunaga during the Iga Rebellion, and when he returned to the site later an attempt was made to assassinate him. The shrine has since been rebuilt. This a sub-shrine to Kōka Saburō, the legendary ancestor of the prominent local family of Mochizuki.

Above: At the conclusion of Toyotomi Hideyoshi's campaign in Kii Province, Kōka District was given as a reward to his general Nakamura Kazuuji. Kazuuji built a new castle here at Minakuchi, from where he enforced the punishment on the Kōka warriors whose negligence had almost caused a disaster during the Kii campaign. The petitions later drawn up on their behalf provided material for the ninja myth.

Below: During the Sekigahara Campaign of 1600 samurai from Kōka distinguished themselves during the defence of the Castle of Fushimi in spite of an act of treachery by a minority whose families had been taken hostage. This is the *ihai* (funerary tablet) in the temple of Jigenji of the Kōka men who died at Fushimi. Their descendants were rewarded by being appointed as hereditary guards at Edo Castle.

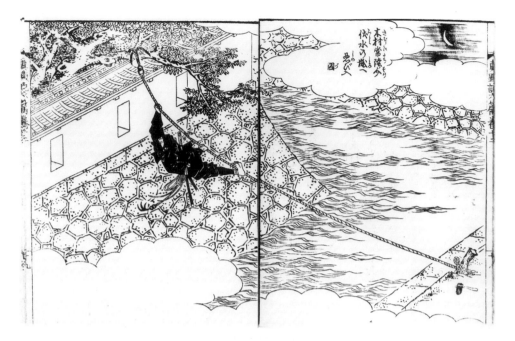

Above: Much of the visual image of the archetypal ninja comes from illustrations of assassination attempts by people who were not the professional infiltrators and spies upon which the ninja myth is based. This famous illustration for *Ehon Taikō ki* shows one such incident: an attempt to murder Hideyoshi in Fushimi Castle by Kimura Hitachi-no-suke. The story in probably fictional but the details of Kimura's attire would provide the stock image of the ninja as it was later developed.

Above: Hattori Hanzō was a general in the service of Tokugawa Ieyasu. He was born in Mikawa Province but was of Iga ancestry, and this connection enable him to assist Ieyasu during the latter's escape through Kōka and Iga in 1582. Ieyasu rewarded the men of Iga by appointing them as hereditary guardsmen at Edo Castle. They carried out investigative work among the *daimyō*, a role which led to the strong association between the term Iga-mono and the activities of spies.

Right: Maeda Toshiie (1538-1599) is reliably recorded as having used an *Iga no nusumi-gumi* (literally a 'band of robbers from Iga') to attack a castle in Noto Province in 1582. This was a year after the destruction of Iga by Oda Nobunaga and strongly supports the theory that refugees from Iga fled to other areas taking their undercover skills with them.

Below: This modern mural painting under the arches of the railway bridge of Okuno Station in the Shinjuku district of Tokyo shows the reality of the work done by the men from Kōka whom Ieyasu took into his service in 1600. They acted as musketeers, not as ninja, and served in this capacity during the Siege of Osaka Castle in 1614-15.

Above: A late, though much-celebrated, act of surveillance by the Shogun's shinobi occurred in 1854 when Commodore Mathew Perry of the United States Navy returned to Japan to demand trade negotiations. Sawamura Jinzaburō Yasusuke, the spy who was sent, was an Iga-mono in every sense of the word. He carried out his work by being included in an official Japanese party for a reception on board Perry's flagship USS *Susquehanna*. He then returned with information. His proud descendants still live in this fine old farmhouse in Iga.

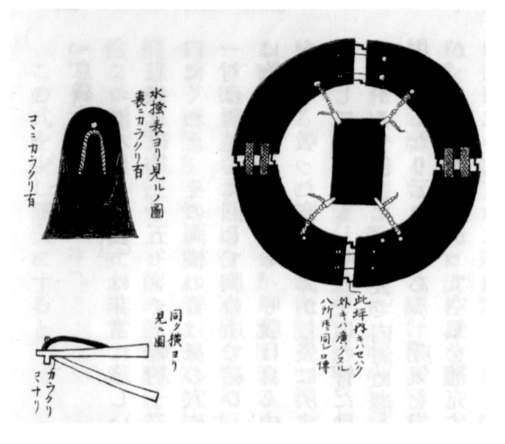

水搔表ヨリ見ルノ圖
表ニカヽフクリ百
コニカヽフクリ百

同ク横ヨリ
見ニ圖

ヽカヽフクリ
ヽナリ

此坪丹キハゼハク
外キハ廣クスル
八所屑同レ口得

Above: Much of what is now regarded as 'ninja lore' derives from an important book of 1676. Called *Mansenshūkai*, it is a compendium of Chinese and Japanese materials concerning espionage and undercover warfare. It includes several weird and wonderful devices of which the most notorious is the water spider. In its original form it was a simple flotation device on which the user would sit.

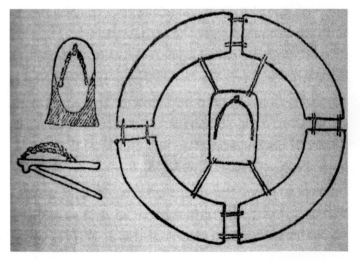

Left: In this illustration from Fujita Seikō's *Ninjutsu Hiroku* of 1936 two separate devices from *Mansenshūkai* have been combined in one to make it into one of a pair of wooden shoes that would allow the wearer to walk on water. This is how the device appears in all modern books about ninja.

Above: The last of the Tokugawa period works on espionage was *Ninjutsu Ōgiden* of about 1800. Unlike most of the others it is very short and is largely concerned with moral and spiritual matters. It contains references to Marishiten, a deity to whom one prayed if one wanted to become invisible. Such a belief among samurai may sound surprising because the war chronicles suggest that once on the battlefield the samurai made every effort to be seen. Yet the invisibility sought from Marishiten could indicate a range of techniques from hiding one's strategic or tactical intentions to the psychological confusion of an enemy.

Left: This important illustration of a *ninsha* from Itō Gingetsu's *Gendaijin no Ninjutsu* of 1937 is probably the first drawing of a ninja figure in a realistic fashion with the overt intent of portraying a specialist of some sort rather than just any warrior or assassin who wished to conceal his identity.

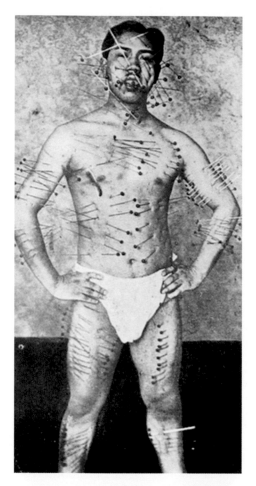

Fujita Seikō shared with Itō Gingetsu a belief that the character *nin* in ninjutsu signified endurance, not secrecy. Fujita took the idea to its extremes through his personal physical austerities, which included eating glass and roof tiles. In this picture he is shown with 250 steel hat pins thrust through his skin.

This illustration from Fujita Seikō's *Ninjutsu Hiroku* of 1936 represents a bizarre blend between the idea of the ninja as a fantasy figure and the realistic portrayal of someone subject to human limitations as would be introduced in the movies of the 1960s. The picture shows the archetypal ninja figure using a modern parachute to jump from a castle wall.

In 1962 a television series called *Onmitsu Kenshi* provided the first true ninja image to be seen on the small screen. At about the same time the film *Shinobi no mono* was released. It was to be hugely influential, but *Onmitsu Kenshi* was aimed at a younger audience and also enjoyed success in a dubbed version in Australia, so it can be argued that its influence was greater.

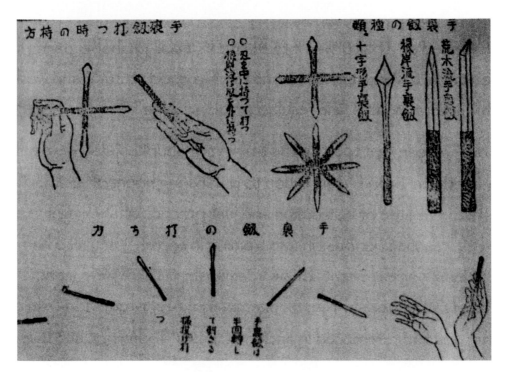

Above: Fujita Seikō's greatest gift to the world of ninjutsu was his introduction of 'ninja stars', the common name given to the cross-shaped and star-shaped spinning iron darts called *shūriken*, to the ninja's armoury. They do not appear in *Mansenshūkai* or any of the other espionage manuals, but Fujita's book includes this drawing based on a *shūriken* owned by the last Tokugawa Shogun. These devices were curios used in the gentlemanly pursuit of the martial arts, but would soon acquire a life of their own as the archetypal ninja weapon.

Right: This roll of ninja toilet paper is probably the most bizarre souvenir item available in Iga City. Okuse Heishichirō, twice Mayor of the former Ueno City, was the pivotal figure in the development of the idea of a ninja. His modest booklet *Ninjutsu* of 1956 was the first modern publication on the subject and set in stone his idea of the ninja as a historical figure subject to human limitations who was indelibly associated with his home town of Iga. Okuse's vision influenced contemporary writers and film producers and laid the foundations for Iga's vast ninja tourist industry.

Left: The Akame Falls in the Nabari district of the old Iga Province have some connection to yamabushi (mountain ascetics) because they are known to have practised austerities such as standing naked under waterfalls. A great conceptual leap has been made to transform yamabushi into ninja, but that is enough to sustain a ninja training centre for tourists where visitors can dress up to throw ninja stars at targets.

Below: The Iga-mono at Edo Castle defended this gate called the Hanzōmon, named after their leader Hattori Hanzō. Its present appearance is little different from this late nineteenth century photograph.

Above: Modern Iga City is the centre of the ninja tourist industry thanks to the efforts of its former Mayor Okuse Heishichirō during the 1950s. One may hire ninja costumes to wear throughout one's visit, and the sight of groups of people dressed from head to foot in black while walking around town raises only the occasional eyebrow.

Below: Kōka's own ninja house dates from the late seventeenth century and was once owned by the Mochizuki family. For many years it was used for the production of pharmaceuticals for which Kōka became renowned. Unlike the Iga ninja house it is also used as a museum with some interesting display cases in addition to the usual revolving doors and secret passages.

Above: One of the most entertaining features at the extensive ninja section in Kyoto's Toei Film Studios Park is this mechanical ninja who climbs along a rope stretched between two castle towers.

Below: One of Japan's newest commercial ninja attractions is the Shinobi no Sato (ninja village), which opened in 2015 in the foothills of Mount Fuji. Unlike older foundations no attempt is made to explain the origins of ninja through museum displays. The focus instead is on entertainment including a 'Ninja Trick House' where visitors make their way through a series of rooms. This picture of a ninja is supposed to give the illusion of a continuing corridor.

same time in works such as Nochi Kagami. It is further evidence for the notion that the invention of the ninja happened gradually over a period of many centuries and that its appearance in the 1950s involved the absorption of old concepts rather than the creation of new ones.

Ninjutsu in pulp fiction and early movies

During the early twentieth century this motley crew of sorcerers and assassins would be amalgamated, transformed into ninja from Iga and Kōka and then set to work using the techniques and tools found in the espionage manuals. The process started within pulp fiction from the 1900s, and much of what a ninja would later do in the movies was anticipated when heroes from the Tokugawa Period reappeared alongside a few new creations within the pages of cheap paperback novels. Many of these influential works were published as part of a series called the Tachikawa Bunko (the Tachikawa pocket books).[8] Almost 200 titles were released by the Osaka publishing house Tachikawa Bunmeidō between 1911 and 1924,[9] and in these lively paperbacks various heroes from the Edo Period lived again 'by sincerity and the sword'.[10] Some were historical characters, others were drawn from the earlier fiction and some were completely new inventions whose exploits would be related so often that they acquired a pseudo-historical reality which they have retained to our day.[11] The Tachikawa Bunko books became a runaway success among the young apprentices who were the initial target audience, and as their popularity spread around the country these adventure stories could not be churned out fast enough. They are still being reprinted in various formats today and have exerted a profound effect on Japanese entertainment. As part of his research into the subject of Japanese popular literature Richard Torrance looked at 62 heroes who featured in the Tachikawa Bunko books and could identify them in 713 prewar movies, 4,349 post-war movies, 618 television programmes and series, and 85 anime, manga or video games.[12]

The most popular character in the series was a hero called Sarutobi Sasuke 猿飛佐助, who was first introduced in Volume 40 in 1914. The original novel about him called *Sarutobi Sasuke* is credited to the author Sekka Sanjin, although this is in fact an assumed name for a team of writers.[13] The cultural historian Adachi Ken'ichi has suggested that the Sasuke stories took their cue from the Chinese adventure novel *Journey to the West*, because Sasuke is so reminiscent of its protagonist, the supernatural monkey king Sun Wukong.[14] Yet to enthusiasts Sarutobi Sasuke has always been a ninja and the 1914 book is commonly regarded as the first 'ninja novel'. For example, in the English-language briefing notes for a European lecture tour on ninja in 2014 by Mie University staff, Yoshimaru Katsuya refers to Sasuke as the 'well-known ninja'.[15]

The reality is somewhat different. The character Sarutobi Sasuke is the son of a samurai who is brought up among the monkeys of the mountains, from whom he learns to jump like them, hence his name Sarutobi (monkey leap). He becomes a retainer of Sanada Yukimura, the *daimyō* of Ueda. To quote Torrance, Sasuke is 'an adolescent superhero who, in addition to his ability to chant incantations, appear and disappear at will, and leap to the top of the highest tree, can hear whispered conversations hundreds of yards away, is superhumanly strong, can ride on clouds, is able to conjure water, fire and wind as well as transform himself into other people and animals'.[16] These are the remarkable attributes that would take Sasuke forward into many future novels and scores of films and television programmes, so it is not surprising that Torrance, looking back from our own age to 1914, is yet another scholar to refer to him simply 'the ninja Sarutobi Sasuke'.[17] It can, however, be quite easily demonstrated that the word ninja does not appear anywhere in the book, because Penn State University Library have recently digitalised their extensive collection of Tachikawa Bunko novels.[18] As the books were written for a young audience, all the ideographs are glossed so there is no ambiguity about any reading. We therefore find descriptions of Sasuke as neither a ninja nor a shinobi, but instead as a 'great name in ninjutsu'.[19] He is also 'an adept at *bujutsu*'.[20] Nor are the words ninja or shinobi no mono used for the mysterious character who appears to him out of a white cloud. Torrance, who translates ninjutsu as 'the ninja arts',[21] calls the stranger 'the foremost master of the ninja arts in Japan',[22] and the man introduces himself to Sasuke as 'a person called Tozawa Haku'unsai, the father of Tozawa Yamashiro-on-Kami, keeper of Hanakuma Castle in Sesshū. For generations in my family my ancestors have taught the secret arts (妙) [known] to the world as the so-called (*iwayuru* 所謂) ninjutsu'.[23]

The 1914 *Sarutobi Sasuke* is therefore not the first 'ninja novel'. It is about a samurai who possesses magical skills along with his martial accomplishments. Yet even if the 'boy wizard' Sarutobi Sasuke was not a ninja in 1914, a particular theme introduced for the first time in *Sarutobi Sasuke* would become very significant in the movies of the future. This was the notion of the hero as underdog, the fighter against oppression and injustice, and it is expressed in *Sarutobi Sasuke* by making him the retainer of Sanada Yukimura, who was defeated by

Tokugawa Ieyasu at Osaka in 1615. Scarcely fifty years before the book was published it would have been unthinkable, and possibly even treasonable, to portray the founder of the Tokugawa regime in a negative light, and in spite of the abolition of the Shogunate such a respectful attitude persisted into the Meiji Period. By 1914 publishers were happy to take the risk, knowing that such an approach would appeal to the humdrum lives of their target audience along with the magic conveyed by the characters' practice of ninjutsu.[24]

The word ninjutsu also appears in the titles of a few silent films from very early in the twentieth century. They were based on the novels, and if these primitive movies are examined in detail it soon becomes clear that the expression ninjutsu still indicates the magic of the Tokugawa Period rather than the martial practices of someone whose behaviour is restricted by his human nature. That concept, which was introduced to action films in the 1960s, would represent a very important shift. The earliest example of a magical ninjutsu movie dates from as far back as 1912 and was about the shape-shifting Jiraiya. The film was called *Jiraiya Gōketsu Monogatari* 児 雷也豪傑物語 (The Tale of Jiraiya the Hero), and sadly no prints have survived. A short film called *Goketsu Jiraiya* 豪傑児雷也 (Hero Jiraiya) followed in 1921 and can be studied in its twenty-minute entirety. It is a classic of the silent screen that shows the heroic Jiraiya as a warrior, sorcerer and performer of magic, even though a recently added set of English subtitles unforgivably translates the title as 'Jiraiya the Ninja'. Jiraiya's shape-shifting is used to great effect in the fight scenes when he changes into a toad and swallows two of his assailants.[25]

Jiraiya's magical victories were realised on the silver screen by special effects and camera trickery, techniques that could also be given the appellation of ninjutsu. Someone who went to see something called a ninjutsu movie in 1921 might therefore be anticipating magical acts produced by trick photography rather than prowess in martial arts, as is confirmed by the amusing fact that when Buster Keaton's 1924 film *Sherlock Jr* was first released in Japan it went under the title of *Ninjutsu Keaton* (忍術キートン).[26] Jiraiya would reappear in several films including *Ninjutsu Jiraiya* 忍術児雷也 of 1955, in which it is somewhat disappointing to note how crudely the shape-shifting is done compared to the 1921 film. Jiraiya changes into and from a toad along with the snake and the slug, but the control wires are clearly visible. Other films during the 1950s would follow the same magical path until the human ninja emerged in 1962, but in the meantime an entirely separate understanding of ninjutsu had arisen, which would influence the ninja myth in a very different way.

Chapter 10

The Enduring Ninja

During the early twentieth century an alternative view of ninjutsu began to emerge alongside the undercover techniques of the military manuals and the magic of the Tachikawa Bunko novels. In complete contrast to the popular understanding of ninjutsu as having something to do with secrecy it took its inspiration from the primary dictionary definition of the character *nin* as endurance and produced an interpretation of ninjutsu related to personal development, much like the emphasis on character building that was being stressed by the founders of martial arts like judo and aikido. The trend also seems to have been related to a desire to subject the topic of ninjutsu to fashionable Western-style enlightened scientific enquiry. Many contemporary scholars of psychology and philosophy were impressed by the scientific approach of the West and attempted to follow the trend within their own disciplines. Ninjutsu was therefore placed under the microscope, although the works that emerged from this new approach were far from being scientific. Instead they brought about an entirely new way of writing about ninjutsu and its practitioners to add to the Tokugawa Period materials and laid further important foundations for the benchmark ninja of the post-war years.

Two names in particular dominate the ninjutsu literature of the early twentieth century. They are Itō Gingetsu 伊藤銀月 (1871–1944) and Fujita Seikō 藤田西湖 (1899–1966), and their important contributions to the ninja myth have much in common. Each begins his key work with regretful comments about how ninjutsu has been misunderstood and misinterpreted as techniques of secrecy. This is a major fault that the new book will put right, but never is this quite achieved. Instead, as will be demonstrated below, both authors would add their own set of errors and misconceptions to the existing corpus of misunderstanding that so annoyed them. That material would enter unchallenged into the received wisdom about ninjutsu and would be repeated endlessly in books about ninja published from the 1960s onwards.

Lying behind Itō and Fujita's ideas is this obsessive concern about the true meaning of the character *nin*. As noted earlier, its primary dictionary definition is endurance, creating compounds that mean patience, perseverance and fortitude, such as the verb *tae shinobu* 耐え 忍ぶ (to endure). It is only the secondary meaning of *nin* as secrecy that produces the expression *shinobi komu* 忍び込む (to secretly enter), and unfortunately (as far as Itō and Fujita are concerned) it is this one that is commonly associated with shinobi and ninjutsu. The tension between the two interpretations lies at the heart of the dilemma facing these influential authors, although neither of them ever seems to make his mind up about which meaning was appropriate for a true understanding of ninjutsu. The situation is not helped by their fanciful extrapolations from the Tokugawa manuals and some imaginative additions to them, which were drawn from the plots of the kabuki theatre and, in one outrageous instance, even a twentieth-century novel. Some of these inventions are so blatant than one is tempted to see them as examples of neither the *nin* in endurance nor the *nin* in secrecy, but rather than *nin* in nincompoop.

The ninjutsu of Itō Gingetsu

Itō Gingetsu worked as a newspaper reporter in Tokyo. He believed that the idea that ninjutsu was about sneaking into buildings by night was a complete misunderstanding that had unfortunately led to an association being made between ninjutsu and criminality.[1] Itō argued that a true *ninjutsu-sha* 忍術者 (his preferred word for a ninjutsu practitioner) depended instead upon the *nin* in *nintai* 忍耐 (endurance), so that ninjutsu involved 'having the ability to maintain one's discipline and focus under the most brutal and challenging of conditions or circumstances… By intense perseverance and effort one's *tanren* 鍛錬 (physical and mental strength) is forged'.[2] Itō extrapolated from this the idea that the practical application of ninjutsu was not infiltration techniques at all, but a form of personal development involving endurance training and extreme physical conditioning. Its goal was an individual's growth rather than his ability to scale walls.

123

Itō's ideas were first aired in his curious, confusing and self-contradictory book *Ninjutsu to Yōjutsu* of 1909. The work begins with an attack on the modernist thought which had led to the criminal activity that pervaded the decadent age in which he was living. Some people, Itō laments, associate ninjutsu with this sort of criminality. However, 'considering how easy it is to commit robbery in this day and age, one hardly needs ninjutsu…'. Having made that point very forcibly he moves on to his thesis of ninjutsu as endurance, and a few pages later he sets out a four-stage training process whereby the goal might be achieved:

1. Training in regulating the respiration
2. Training in making the body move lightly and being able to walk rapidly
3. Training in overcoming unforeseen circumstance or changing plans and enduring difficulties
4. Training in *bugei*, *jūjutsu* and other technical training.[3]

Over the next thirty-five pages the reader is taken through the four steps. The greatest emphasis is laid upon the breathing techniques found in Stage One. Stage Two includes the famous ninjutsu technique of walking sideways by crossing one leg over the other, and Itō's book is almost certainly the first time that this idea appeared in print. There is an accompanying drawing. It is quite crude, but shows the *ninjutsu-sha* dressed in black and walking in this fashion. The figure resembles the Tokugawa Period book illustrations noted earlier, but with one big difference: the man is neither a named assassin nor a symbolic invisible man in a theatre. He is instead a specialist in a non-magical form of ninjutsu, so Itō has now provided for the first time the one key visual image that was missing entirely from the military manuals.[4] Stage Three expands upon his notion of endurance and perseverance. Martial arts appear only in a very poor fourth place and should only be undertaken by the trainee *ninjutsu-sha* if he has mastered the vitally important breathing techniques, because the practical side of ninjutsu as Itō envisages it is so different from conventional martial arts training. It is also important to note that Itō does not regard ninjutsu as a martial art in itself.

All this is very convincing, but Itō then surprises the reader by totally contradicting himself, because in the next chapter he describes his *ninjutsu-sha* as a master of stealth who climbs into defended places. It is a remarkable U-turn, and his account includes two classic concepts lifted straight from the Tokugawa espionage literature and only slightly elaborated. The first is the *ninjutsu-sha*'s clothing, which should be black tinged with red. The second is the use of the ninja sword for scaling walls. The weapon should be long enough to allow one to use the sword guard as a step, so Itō's perfectly trained exemplar of human endurance does indeed climb walls after all![5]

The reader is given a further shock when the page is turned and the topic changes to *yōjutsu* (magic techniques or sorcery), beginning with a chapter entitled, 'The Technique of Turning Into a Rat'. The book continues with many pages about the real-life art of making oneself invisible that can be summarised in the two words: 'blend in'. Strangely, the examples he chooses are all drawn from works of fiction about magical slugs and toads, but the reader is given the impression that Itō actually believes in the reality of such techniques, unless he is using very heavy-handed irony:

> ...if one were to take Ninjutsu, the training method for forging the mind and the body, and drill and condition yourself relentlessly, beyond even that which would be required for Ninjutsu, the effect would be dramatic. Without twisting the fingers into patterns and without uttering curses or spells you would naturally be able to turn yourself into a toad. Transform into a snake. Form yourself into a slug.[6]

These bewildering passages beg the question as to Itō's sources. Happily, he would eventually reveal them to the bemused reader in his follow-up book of 1917 called *Ninjutsu no Gokui*.[7] Itō explains there that he was taught many things personally by a certain Tanimura Ihachirō, a man from the former Echigo Province, whose conversations provided much material and many 'hints' for future research. In second place are the many books and scraps of information he has gathered from elsewhere. Itō cites the military manual *Shōninki* as his third source. His copy was not obtained from some sage-like ninjutsu master living in the fastness of high mountains but, he reveals, by the mundane process of visiting Ueno City Library.[8] Itō appears to have used a handwritten version and was intrigued by its 'troublesome' references to material that is only to be transmitted orally. It is interesting to see use being made of manuscripts held by the former Ueno City (now Iga City) Library long before Iga's ninja boom had begun, and it is also noticeable that Itō makes no reference in any of his works to *Mansenshūkai*. As Yamada notes in his comments on Itō's book, *Mansenshūkai* is nowadays regarded as the 'kernel' of ninjutsu studies, yet Itō does not mention reading it.[9]

Itō may have made no use of *Mansenshūkai*, but he also made little use of *Shōninki*. It is not mentioned in the introduction to *Ninjutsu no Gokui* where he revisits his early ideas about the true meaning of ninjutsu. Toads and slugs have thankfully been left behind as he returns to his belief that because of the newly introduced fad of motion pictures an erroneous idea of a *ninjutsu-sha* has been widely embraced by people who are sadly ignorant of the truth.[10] He later repeats his earlier dismissal of this understanding of ninjutsu as 'a mysterious method of fighting... involving sorcery of some kind... the so-called Ninjutsu in moving pictures and story books'.[11] Itō wishes to put his readers right and in between the above comments he presents a history of ninjutsu in which almost anyone in Japanese history can be labelled a *ninjutsu-sha*. His sources for this section include passages from war chronicles, but he relies above all on *Buke Myōmokushō* of 1806. This is of course a work from the later Tokugawa Period in which the ninja myth has already entered a highly developed phase, but Itō erroneously regards it as a primary source which provides proof for his notions of historical ninjutsu.

Once this questionable historical section has been concluded, Itō again contradicts himself by presenting yet another definition of ninjutsu as a general term for those who use a variety of methods to condition both mind and body 'for the purpose of observation and infiltration'.[12] By including those words Itō links his ideas about ninjutsu to the secret entry techniques that he both dismissed and then accepted eleven years earlier. This is typical of the confusion and repetition found in his works. He then goes into more detail about the practical techniques of espionage and infiltration. *Shōninki*, and possibly even the unacknowledged *Mansenshūkai*, are his likely sources. There is, however, very little that has not been covered already in *Ninjutsu to Yōjutsu*. Once again he mentions wall climbing using a sword with an extra-large guard, and describes the preferred tone for a ninja's clothing, which is discussed in terms of the newly introduced practice of military camouflage.[13] The latter half of the book is given over to a long-winded exposition of escape techniques taken straight from the Tokugawa literature.

Itō Gingetsu would return to the topic of ninjutsu for the last time in his *Gendaijin no Ninjutsu* of 1937.[14] Ninjutsu is now simply defined as 'surveillance techniques', for which the training in ancient times was violent and painful. For the *gendaijin*, the 'person of modern times' to whom the book is addressed, the practical application of ninjutsu begins not with endurance, but with techniques of self-defence against an assailant, and Itō supplies several alarming suggestions involving improvised defensive objects such as lighted cigarettes and fence posts.[15] He then writes about the ninjutsu of the past. To 'eliminate the colour of one's body' the *ninjutsu-sha* wore suitably dark clothing 'that melts into the darkness of night'.[16] Yet the *ninjutsu-sha* also had to conceal his form, and crucial to achieving this is once again the technique of walking sideways, for which a helpful picture is included. It shows the classic ninja image of a man in black, and shortly afterwards there is a three-part illustration of a similarly dressed warrior (here identified as a 忍者 but glossed as a *ninsha*) taking on a swordsman.[17] These beautifully drawn pictures are very similar to the *Ehon Taikō ki* picture and contrast with his rough 1909 illustration. They are almost certainly the first time the figure in black had ever been included in any work in a realistic fashion, with the overt intent of portraying a specialist of some sort rather than just any warrior or assassin who wished to conceal his identity. They had in fact been preceded by one year in a book by Fujita Seikō, as will be discussed below, but Fujita's picture was of a fantasy figure, so in the form of his *ninsha* Itō has given us our first developed image of what would become the benchmark ninja.

Itō concludes his book with a jarringly sharp modern reflection. It is now 1937 and Itō believes that ninjutsu has a role to play in modern warfare. Japan is under threat and, 'Hitlers and Mussolinis appear here and there and are gone just as quickly. We can hardly postulate the theory that, "They won't come here"'.[18] The idea is sadly not developed, but it is interesting to note that in 1937 Itō perceives the two dictators as potential enemies rather than the allies they would become.

Commenting upon Itō's work, Yamada recognises Itō's desire to study ninjutsu scientifically.[19] Yet in this Itō fails utterly, because his books are lacking in anything that might be called scientific rigour, and his confused and sometimes baffling style would have made them almost inaccessible to the general reader. As for the techniques of ninjutsu, all we get is a repetition of Tokugawa Period material about walking sideways, the ninja sword and the black costume. Apart from this the only real developments of the ninja myth to be found in Itō's works are the illustrations of the *ninsha* in *Gendaijin no Ninjutsu*. The ninja was beginning to take shape, but he still had a long way to go.

The ninjutsu of Fujita Seikō

Itō Gingetsu's books may have described various sorts of extreme ascetic practices that were part and parcel of his understanding of what ninjutsu was, yet he never refers to undertaking them personally. In this he differed from the other highly influential figure in pre-war ninjutsu studies: the remarkable Fujita Seikō. In his writings, and even more through his own practice of bizarre physical austerities, Fujita took the *nin* of endurance to its extreme, and when Japan entered a new world after World War II he returned to the subject and would become instrumental in creating the ninja of today.

Fujita Seikō was the son of a Tokyo policeman whose skills at the martial arts had not been confined to any *dōjō*. Instead Fujita's father had become something of a hero for tackling armed criminals using techniques which he would pass on to his son, although many claims about Fujita's early life, military service and martial arts training appear to have been greatly exaggerated.[20] For example, Fujita's grandfather, who taught the boy ninjutsu, claimed to be the descendant of Wada Koremasa, the historical figure noted earlier in this book as a leading *kokujin* of Kōka who gave refuge to the fleeing Ashikaga Yoshiaki. Koremasa was believed by the Fujita family to have founded a school of ninjutsu called the Wada-ha Kōka-ryū, an organisation with its own *hiden* tradition into which young Fujita Seikō was to be initiated. All this is highly questionable, because the historical record has Wada Koremasa leaving Kōka in 1568. Nevertheless, following Fujita's intensive training his dying grandfather passed on to him the headship of the Wada-ha Kōka-ryū and gave him some ancient scrolls that are believed to have included *Ninjutsu Ōgiden*.

Fujita Seikō's claim to have inherited a particular tradition of ninjutsu is of less relevance to this book than his overall understanding of it, which is very well recorded and begins with his belief in physical conditioning and training along the lines of Itō's four-stage programme. There can be little doubt that Fujita was highly skilled in various martial arts and underwent physical training of great intensity, but his idea of ninjutsu involved heightened sensory awareness that went far beyond any of Itō's admonitions and examples. The obsessive Fujita Seikō believed in a control of pain and bodily functions that is more reminiscent of yoga and occultism than martial arts training, although just like Itō Fujita he does not seem to have quite made up his mind about the relationship between *nin* as endurance and *nin* as secrecy.

Both understandings of *nin* receive attention in his important book *Ninjutsu Hiroku* 忍術秘録, which was published in 1936. The great puzzle about *Ninjutsu Hiroku* is why someone who had supposedly been initiated into a mastership of the secret arts (something that never happened to Itō Gingetsu) should wish to reveal them through the printed word, but Fujita is quite frank about his motivation for producing the book. It is nothing less than a patriotic act that will help Japan in its present war against China. In a short introduction he reminds his readers that 'Ancient ninjutsu is called a secret art, so the resulting literature is scarce and much was transmitted from master to disciple by means of oral tradition'. However, he now has in his possession thirty different works on the subject contained within fifty volumes in all, which he intends to publish and transmit to a new generation. 'I am the only person nowadays who can pass on ninjutsu,' says Fujita, and he will do so to the whole country as he has already done in a more restricted way through his lectures and demonstrations. *Ninjutsu Hiroku* will 'plainly reveal the secrets of ninjutsu', and his motivation is clearly the good of the nation. He quotes the phrase once uttered by the famous *daimyō* Hōjō Sōun that, 'a general's need is invariably to win the hearts of heroes'. The art of war has to be understood in the present day, and he ends on an optimistic note that, 'therefore by means of this book I believe with a broad smile that "Ninjutsu will now be understood"'.[21]

Ninjutsu Hiroku represents an important stage in the creation of the ninja myth. The military manuals of the Tokugawa Period, rather than any secret wisdom passed on from his grandfather, appear to have been Fujita's main sources, and all the well-known ones are listed in his bibliography including *Mansenshūkai*. No one before Fujita seems to have made any use of it, so Fujita must be regarded as the man who 'discovered' *Mansenshūkai*, and he employs it extensively. Much of the material in *Ninjutsu Hiroku* about the *nin* of secrecy is in fact lifted straight from *Mansenshūkai* with almost no rewriting, although his first reference to material from any Tokugawa manual is to *Shōninki*'s classic list of seven disguises. He also mentions rats, snakes and toads, but unlike Itō's mystical ramblings about animal transformations Fujita's use of the concepts comes over as very sensible, even though his illustrative examples are drawn from kabuki plays. For example, if the infiltrator imitates the sound of a rat the credulous guards might think he has changed into a rat. Otherwise a disturbed pack of dogs causes such confusion that the shinobi easily escapes, and the constant theme of escape in the early part of the book reflects *Mansenshūkai's* admonition that the shinobi's role is to return alive with information. Even Fujita's account of walking sideways makes perfect sense because it is the technique the shinobi would have used outside a gate or a wall. He would stand with his back to the obstacle and slide sideways to avoid detection. Techniques like these, he says, which could give the impression of invisibility, would be understood as something magical centuries ago, but in reality they involved logical scientific thought.

At this stage in his book Fujita is writing with great clarity, and he also notes the paradox in *Mansenshūkai* that even though the shinobi has to return alive with information, he must cast away all desire to live. When it comes to the techniques of infiltration and information-gathering *Mansenshūkai* is again Fujita's main source, and just like *Mansenshūkai Ninjutsu Hiroku* contains quotations from Sun Zi. His discussion of *yōnin* and *innin* is also drawn wholesale from *Mansenshūkai*. There are sections on how to probe a room using a sword and its scabbard, how to recognise patterns of sleep, why back door entry is preferred to the front door and the advisability of sneaking into a house while there is a celebration going on. The main interpretive comments Fujita adds to the *Mansenshūkai* material concern its application to modern warfare, which make his book a fascinating read. Escape techniques using water, smoke, stones, vegetation and other forms of cover have genuine analogies with modern conflict. Stones and water can be thrown in a pursuer's face, and a fleeing spy can set fire to grass so that it spreads towards the enemy 'like poison gas'. The observation of how fish move is useful for escaping through water, and Fujita compares their streamlined shape to airships and submarines.

The tone of *Ninjutsu Hiroku* changes quite abruptly after these straightforward sections because the *nin* of secrecy then gives way to the *nin* of endurance as it was practised by Fujita Seikō himself. 'The *nin* of ninjutsu is the same as the *nin* of *nintai*' writes Fujita, and with that he launches into a long discussion of his own austerities and physical prowess. Over the next few chapters *Ninjutsu Hiroku* becomes largely an exercise in self-glorification, because Fujita's main interest is his own idiosyncratic interpretation of ninjutsu rather than the activities of ancient spies. There is a photograph of him walking on his toes with his feet bent inwards. He also mentions breaking bottles over his head (Fujita's record is forty) and the technique of driving nails into wood using one's forehead as a hammer, which he does not claim to have tried. A skilled shinobi, according to Fujita, can even avoid being poisoned by training his body to eat anything, even glass or a roof tile, which he is able to do. The reason for carrying out such unpleasant activities, he states, is so that the shinobi can withstand torture, and he includes a photograph of himself with 250 hat-pins stuck through his skin.[22] He was evidently proud of stunts like these and talked about them in an interview he gave to *Newsweek* in 1964, two years before his death. The anonymous journalist, who describes Fujita somewhat unkindly as 'an occasional throwback', quotes him thus: 'to learn to tolerate poison I ate sulphuric acid, rat poison, wall lizards, 879 glasses, 30 bricks. A glass was easy but it took me 40 minutes for a brick'.[23] In *Ninjutsu Hiroku* Fujita adds that he is also a skilled mimic of animal sounds and musical instruments. Just like the shinobi in *Shōninki* he is proficient in any guise he might choose to adopt, so (of course) he also dances very well.

Fujita's claims would probably be dismissed as fantasy were it not for the photographic evidence, but there also exists an independent eyewitness account of Fujita in action, written by the celebrated English potter Bernard Leach. In *Beyond East and West* Leach includes an account of a meeting in Japan with a mysterious 'Mr X' who ate glass and pierced his skin. Part of the performance was filmed and a viewing of the footage confirms that the man was indeed Fujita Seikō:[24] Leach described the event as follows:

> It was about this time that Hamada reminded me of a very strange man of whom he had often spoken at St Ives in the old days. This Mr X had been trained from youth in the manner formerly employed for traditional spies – highly skilled men who carried secret messages or obtained information for their feudal lords... Mr X agreed to come to my house and on a fine afternoon talked to some fifteen of us for two hours. No other foreigner was present and my guests were mostly literary men. Mr X said that he began his training at the age of six when he was selected from other child candidates. There were two tests; one was silent crossing of a room covered with crumpled paper, the other was to keep his head under water for two minutes.
>
> Mr. X was a short, rather tubby man, but when after talking for a quarter of an hour or so describing his further training, he threw off his kimono and stripped to his drawers, we could see a very neat, well developed body – when he drew in and extended his abdominal muscles the range was quite unusual. The first thing he

did was to light a cigarette, put the butt end into one nostril and with a finger on the other, inhale deeply. Then he drank from a glass of water and only subsequently exhaled. This slight beginning rather astonished some of us, even though Hamada had told us long ago that this man had cultivated quite extraordinary abilities, such as climbing walls at right angles without foot- or hand-holds, by using arm- and leg-pressure against each surface.

The next thing he did was to take some women's Victorian six-inch long hat-pins with black or white glass knobs at one end. After wiping them with alcohol he opened his mouth wide and slowly pushed first one right through his cheek from the inside in one direction and then another in the opposite direction.... Mr X even went on talking with these hat-pins protruding and exhibiting no signs of either pain or bleeding. Then he calmly withdrew these weapons slowly, thrust forth his tongue and holding the end with one hand, perched a third hat pin through it, the point emerging on top. Thus he went on speaking, then after withdrawing it he requested another tumbler, which I provided. He examined it, then put it well into his open mouth and bit out a piece at least two by one inch which he proceeded to crunch between his jaws for a minute or so, then gulped down a mouthful of water from that first glass, then went on talking normally. This was astounding! There he stood alone amongst us, and I have this visible record as well as the memory of hearing further crunching of small remnants of glass whilst he went on lecturing.

Anyone might ask, 'What was the use of such revolting tricks?' I assure them there was no trickery, and the use of them saves life or puts off suspicion when in a tight corner. Between the age of six and twenty this man had been trained to jump six feet in height and over twenty feet in length. He could run or swim twenty miles; he could eat poison and suffer no harm, get drunk on water and remain sober after a couple of pints of spirits – so my friends told me. The knowledge of dogs' language enabled him to call them at any time and encourage them to fight themselves.

What I saw him do next was to gather in one hand the rings from a five-to ten-pound steelyard weights totalling about seventy pounds and repeatedly throw them at arm's length in the air and crash them on his chest. There was no mark or sign of pain. His explanation was that though the training was part physical it was mainly mind over matter. One might ask, 'What is its limit?'.[25]

Even though stunts like these were an expression of his understanding of the *nin* of endurance, in *Ninjutsu Hiroku* Fujita keeps coming back to the *nin* of secrecy, for which his sources are twofold: *Mansenshūkai* and his own dogged creativity. In addition to the material noted above *Mansenshūkai* provides information about survival rations, the ninja sword and caltrops, while *Shōninki* adds the six tools. The famous hand signs are clearly illustrated, as are a mixture of other *Mansenshūkai* devices and some police weapons such as the sword-catching *jitte*, sticks and knives. The practical section of *Ninjutsu Hiroku* concludes with these weapons, but of much more relevance to the ninja myth is the use Fujita makes of his own vivid imagination, for which he offers no apology. To Fujita ninjutsu is a living art that will benefit Japan in its current war, so historical material can be freely interpreted, modified and developed to face the new challenge. The result of this process was the transformation of some fairly mundane ideas and objects from *Mansenshūkai* and other unacknowledged sources into fantastic concepts, tools and weapons that are nowadays regarded as historically authentic.

Fujita's creative process begins in some style when he states that the shinobi no mono of the past could jump from a great height using the towel from his 'six tools' as a makeshift parachute. Not content with making this nonsensical claim, Fujita illustrates it with his own drawing of a shinobi jumping off a castle wall wearing a modern parachute on his back. Apart from its sheer audacity, the most interesting thing about this picture (which predates Itō's drawings of the *ninsha* by one year) is that the figure is shown in the now accepted costume of the benchmark ninja. That makes it the first realistic drawing of the ninja we know today, although the inclusion of the parachute places it in the category of magical fantasy rather than the human figure.

The section continues with a recommended training programme for jumping skills, which was to plant fast-growing hemp and leap over the plants as they grew taller during the year. It is all very entertaining, but the idea also appears in a novel called *Tenpei Dōji* by Yoshikawa Eiji, which was published at about the same time as *Ninjutsu Hiroku*. A fantasy warrior is rescued as a child and as part of his training is made to jump over fast-growing hemp for 120 days.[26] It is impossible to ascertain who thought of the idea first, but the idea of it being part of traditional ninja training entered the vocabulary and even attracted the attention of the famous violin teacher Dr Suzuki Shinichi. 'Long ago I read in a book whose title I no longer recall that the training method for

133

ninja in the old days included the exercise of learning the high jump: "Plant a young hemp, and jump over it every day"'. Suzuki felt that it was an excellent analogy with his insistence on daily practice for his young violinists.[27]

Ninjutsu Hiroku also includes a strange picture of a water spider. It is clearly based on the one in *Mansenshūkai* and next to it is a drawing of one of the hinged *geta* (clogs) used by farmers for walking through waterlogged ground that also appears in *Mansenshūkai*. For some unstated reason Fujita combines the two pictures into one, so that a *geta* has been added to the centre section of the water spider! There is no accompanying text, so once again the reader is left puzzling over Fujita's sources and intentions.[28] The brief caption states that it was used for 'walking in water' not 'walking on water', so it is not quite the device that is seen in ninja theme parks today. Nevertheless, Fujita's depiction of its centre section as something on which you would place your foot rather than your backside very probably represents the moment of creation of one of the ninja world's most notorious inventions.

Fujita Seikō's greatest contribution to the ninja myth was his treatment of the *shūriken*, a topic so important in the invention of the ninja tradition that it will be given a later chapter all to itself. Through devices like these Fujita Seikō holds a unique place in the development of the ninja myth, because his inventions provided a bridge between the world of *Mansenshūkai* and the post-war ninja of the movies. The parachute may have been just too ridiculous to survive, but *shūriken*, hemp-jumping and the water spider have been largely accepted as historical fact. Fujita's imagination therefore complemented the creativity of the Tachikawa Bunko authors and further reinforces my overall thesis that the ninja myth has involved a centuries-long process of a constant reinvention of tradition.

Ninjutsu in World War II

This chapter will conclude with an examination of Fujita Seikō's wartime role. Like Itō Gingetsu, Fujita Seikō lived at a particular moment in Japanese history when ninjutsu had to be considered as a possibly useful tool for modern warfare. Unlike Itō, however, Fujita had a chance to turn these ideas into reality, although he would be cruelly disappointed. Fujita Seikō's military service in World War II involved training intelligence officers, and in his book of 1942 *Ninjutsu kara supaisen e* (From ninjutsu to spy warfare) he explained how ninjutsu could be of benefit in the war effort. This book incorporates much of

Ninjutsu Hiroku, with the addition of chapters that are almost totally orientated towards the contemporary situation.[29] Spies nowadays, Fujita warns, come disguised as military attachés or embassy officials, and espionage is something practised by every country. Ninjutsu is mentioned in a historical context, with several examples of historical practitioners during the Sengoku Period, but this material is always subordinated to the needs of the times, and instead of using *Mansenshūkai* as a source of examples he relates anecdotes about spies during World War I. Secret messages can be conveyed by using them to make roll-up cigarettes, he says, so that if a spy is discovered he can smoke the evidence. Otherwise there is a warning similar to the British government's famous wartime admonition that, 'careless talk costs lives'. The only time he concentrates on the ninjutsu of the olden days is in his statement that the secret operative lives 'with his heart under the edge of a blade', a reference to the *nin* ideograph. Just as in ancient times the modern shinobi goes into enemy territory prepared to throw away his life in pursuit of his mission, but while others might be satisfied with giving their lives, the shinobi cannot afford to do this. He must return with information.[30]

Fujita later recorded that his attempts to incorporate elements of ninjutsu training into the curriculum failed through of lack of time because the Japanese army needed men who could be trained quickly and sent off immediately into battle.[31] As for any wider dissemination of Fujita's ideas among the general public, the 1940s were hardly an ideal time for the start of a ninjutsu craze, and by then Fujita's own emphasis was more on the martial arts and how they might be applied to the wartime situation. He condemns the popularity of Western sports in Japan, of which baseball is his *bête noire*. There should be no reporting of baseball games in newspapers. If it is ignored, the alien import will wither away due to lack of interest. 'Abandon sports and return to *bujutsu*' is Fujita's rallying cry. Japanese martial arts are what young people should be practising, and it should be full-contact. 'We must be prepared for blood to be shed in these matches. In fact since it is easy to do shallow cuts and cause bleeding with the shinai (the bamboo sword) several times before what would be a fatal blow would be made, we should prepare for a lot of blood to cover the floor at the end'.[32]

The Japanese government agreed entirely on this point. Not only should Japanese martial arts replace Western sports, but *bujutsu* also had to change to a less sports-orientated and more militaristic practice, so that full contact kendō could contribute as much to the martial spirit of the nation as the modern 'martial arts' of bayoneting and grenade-throwing. Kendō in its sporting form was apparently dismissed as 'dancing with bamboo poles'.[33] Instead, the art of the Japanese sword was presented as a new cult of cold steel, against which an enemy would be petrified. In his detailed study of the transformation of kendō during the war Bennett shows how it was a 'return to kenjutsu, where the aim in battle was to kill, and outside of battle was to provide the Japanese people with an emotional bond to a warrior past'.[34]

Through initiatives such as these, tens of thousand of young men were exhorted to embrace a very narrow interpretation of *bushidō* and imitate the samurai of old by their patriotic loyalty. In a cynical twist certain heroes and motifs from samurai history were used to add colour and nostalgia to the dark reality of the soldier's likely death. Thus in the Yūshūkan, the highly nationalistic military museum attached to the Yasukuni Shrine in Tokyo, one can see a suicide plane with a cherry blossom painted on its fuselage as an emblem of the samurai's brief life. Next to the plane is a suicide submarine that bears upon the conning tower the family crest of the doomed loyalist samurai Kusunoki Masashige.

Japan's samurai heroes were therefore pressed into service for the war effort, but it is important to note that the list did not include any practitioners of ninjutsu. No kamikaze mission was ever compared to the surprise attack of a shinobi, and no suicide plane ever took off with Itō's classic *ninsha* image painted on its side. The total absence of the visual imagery of ninjutsu from wartime propaganda shows that in spite of the popularity of the Tachikawa Bunko novels, the topics played no role in the wartime consciousness of the general population. The lack of any mention of them in patriotic exhortations during World War II, a time when such heroics would have been most useful, also confirms beyond any shadow of doubt that the idea of the ninja as an easily recognisable military icon is completely a post-war phenomenon. The black-clad ninja, who is now such a powerful martial symbol, was the product not of a time of war, but of the ensuing time of peace.

Chapter 11

The Shinobi Awakes

As Japan emerged from the horrors of war and the brief humiliation of the Occupation, the ninja myth experienced its definitive transformation. From small beginnings during the 1950s it reached its fruition in motion pictures to give birth to the ninja as we know him today and to fix for evermore the accepted reading of his name. The end result is very well known; the means to it much less so, and in two specific aspects the invention of the ninja happened in reverse to what is usually understood. First, the creation of the ninja image by the movie industry in the 1960s did not merely add fantastic elements to an existing historical character. To a very large extent it removed the fantasy from existing fictional characters to make them more human. Second, long before the first ninja movie was released another surprising development had already taken place within the world of tourism. These two developmental strands of fiction and tourism overlap in time, but are best dealt with as separate issues, so the novels of the 1950s will be covered later in this chapter. We will begin with the surprising emergence of a fledgling ninja tourist industry at the beginning of the decade. This contrasts with the usual assumption that the ninja of the movies kick-started ninja tourism, because a highly localised form of it had already been in existence in the Iga area for over ten years before the first ninja movie was ever mooted.

The vision of Okuse Heishichirō
In a previous chapter I likened the Iga ninja tradition to Sleeping Beauty because it had lain dormant since the Tōdō family's takeover of the territory in 1608. In 1947 a handsome prince arrived to awaken her with a kiss. Iga's Prince Charming was a remarkable man called

Okuse Heishichirō 奥瀬平七郎 (1911–97), who would lay the foundations for the definitive ninja image and its now unquestioned link to the Iga area. Okuse was a native of Iga and became Head of the Planning Department in what was then Ueno City Hall in 1947. He would go on to become the city's mayor and serve two terms of office. A reading of *Mansenshūkai* had made him realise how its connection with Iga could be used to boost local tourism. With that one moment of inspiration Okuse Heishichirō single-handedly invented the ninja tourist industry and, arguably, the ninja himself, because no one in modern times has had more influence on the image and mythology of the man in black. Okuse's vision influenced the ideas that were being developed by contemporary writers and illustrators and also anticipated by a good fifteen years the decisive moves of the film producers with whom he would work to transform his ideas into the ninja of today, a process in which his close relationship with Iga would be a crucial factor.

Okuse Heishichirō's efforts on behalf of his home town began in 1951. A children's exhibition was to be mounted in Ueno City, so Okuse incorporated into it a 'Ninjutsu Wonder Museum' (忍術不思議館) where weapons and secret writings were displayed for the first time as a commercial attraction. Among those who helped Okuse realise his vision was Fujita Seikō. He supplied some of the exhibits, and it may well have been Fujita who led Okuse to read *Mansenshūkai* in the first place. Throughout the war and the Occupation's ban on the martial arts Fujita had been one of the few men who kept alive their memory and practice, and it was because of this, rather than his skills at ninjutsu, that he was to be honoured and revered in post-war Japan. Indeed, once peace was restored he seems to have been reluctant to practise ninjutsu openly, saying that it was a dangerous anachronism in the modern world.[1] Nevertheless, he obviously approved of the idea of ninjutsu being valued as a unique cultural asset. He had said as much in *Ninjutsu Hiroku* in 1936 and now willingly assisted Okuse on the display's preparation and gave demonstrations of ninjutsu during its run. It was the beginning of a hugely rewarding partnership. The children's display attracted young and old alike, and when interested tourists arrived in Ueno City well after the exhibition had closed Okuse gave lectures about ninjutsu and members of staff dressed in black to put on ninjutsu demonstrations.

The ninjutsu equipment had also proved to be so interesting to visitors to the Ninjutsu Wonder Museum that a permanent display area was sought. Iga already had a museum devoted to its historical samurai past in the shape of the Igagoe Historical Materials Hall, which commemorated a famous vendetta in 1634, so in 1961 Okuse created a Ninjutsu-kan 忍術館 (Ninjutsu Hall) inside it. In 1964 a farmhouse said to have belonged to an Iga shinobi called Takayama Tarōjirō was transferred in sections to the grounds of Iga-Ueno Castle to become Japan's first 'authentic ninja house'. The museum collection was displayed within the ninja house until today's purpose-built Iga-ryū Ninja Museum was opened in 1970. Okuse also went round Iga arranging for monuments to be erected to local heroes such as Hattori Hanzō, and marking with noticeboards and standing stones the sites of castles and battles from the Iga Rebellion. In 1963, while the ninja movie craze was only just developing, Iga's annual Cherry Blossom Festival was replaced by a Ninjutsu Festival. It was discontinued in 1969, but resurrected in 1979 as the Ninja Festival, which continues in a modified form to this day to provide an annual stimulus for the never-ending celebration of Iga's fortunate heritage.

Okuse Heishichirō also became very influential through his writings. His books would become the major source for Andy Adams' innovative articles about ninjutsu in *Black Belt* in 1966 and 1967 and Adams' misleadingly titled book *Ninja: the invisible assassins* (1970). This meant that along with his influence in the film world, Okuse's ideas reached the West long before most other writers on the subject, and because Okuse's views were backed up by the physical existence of the museum objects, his interpretations and above all his Iga-centric view of ninja became the accepted norm. Through his devotion to Iga Okuse therefore resembles Kikuoka Jōgen, who had celebrated a local tradition in *Iran ki* while simultaneously shaping it. Yet lying behind all of Okuse's works is his idea of the ninja as a tourist attraction, and it is probably for this reason that he tends to be ignored by serious modern writers on the subject. His name appears nowhere in the recent books by Yamada, Kawakami or Nakashima, all of whom recognise the work of Itō and Fujita, yet studiously ignore the man whose contribution to the ninja myth (and their own careers) was absolutely fundamental.

Okuse Heishichirō went into print in 1956 with the first post-war publication on the subject of ninja. It was a simple 26-page booklet entitled *Ninjutsu*, which was published by a local railway company, and its intended role in tourism promotion is illustrated by the inclusion of a map showing how the ninja-related sights could be accessed by rail. Okuse Heishichirō's name appears on the title page, although it is now accepted that the novelist and cultural historian Adachi Ken'ichi helped put his ideas into words. The two men were close collaborators and both would go on to produce the first full-length books about ninja and ninjutsu in the post-war period, in which material was reused from this modest yet remarkable joint effort.[2]

Ninjutsu is a unique piece of work in which much of what we now accept as the benchmark ninja is set out in print and photographs for the first time. Gone are the incoherent ramblings of Itō Gingetsu and the painful physical austerities of Fujita Seikō. Instead Okuse follows Fujita in his employment of *Mansenshūkai*, but totally transforms the use Fujita made of it, even though some of his interpretations are almost as creative as those of his illustrious predecessor.

Ninjutsu's greatest contribution to the ninja myth was its innovative repackaging of the topic. There is only a brief mention of ninjutsu as magic and the booklet is certainly not a sombre tome about ninjutsu as part of the war effort. Instead it represents ninjutsu for a new post-war generation, expressed in the vivid context of the 'authentic' ninja sites of Iga and the city's collection of 'authentic' ninja equipment, all of which would be illustrated photographically for the first time. The result was the creation of a highly convincing world for the ninja to live in, and in spite of the booklet's title, *Ninjutsu* represents a major shift in emphasis from ninjutsu to the ninja himself. The young reader who had enjoyed the 1951 exhibition and had gone on to buy the booklet would be given the impression that Iga had ninjutsu for its inheritance, a concept that was exciting and authentic. In its attractive layout, its likely use of the word ninja (it is sadly unglossed) for the first time in a non-fiction book, its concise yet exciting contents and particularly through its photographs *Ninjutsu* is the forerunner of every ninja book that would follow in the 1960s.

Ninjutsu begins, fittingly enough, with Okuse Heishichirō's 'dream', which echoes the earlier despair of Itō and Fujita that the concept of ninjutsu was misunderstood:

> The fantasy idea of ninjutsu as being to do with an individual of superhuman strength is not confined to children. From my point of view ninjutsu is a dream that somehow expresses people's desires. It is not only found in Japan, because we have the Monkey King and the Thief of Baghdad carried away on a flying carpet. I could give many examples of such stories. So, how can we describe my dream of ninjutsu? What actual form did the original ninjutsu take?

On the opposite page from this quotation is a picture that sums up the fantasy notion that *Ninjutsu* will soon leave far behind. It is of a film actor playing the role of Sarutobi Sasuke in the 1953 movie *Ehon Sarutobi Sasuke*. Okuse appends a quotation from the original novel about Sarutobi Sasuke representing a 'Don Quixote-type' of ninjutsu. From this Okuse moves rapidly on from magicians to his idea of a ninja who is subject to human limitations, a notion that would soon become the conceptual norm. Using photographic evidence to back up his ideas, he begins with a picture and description of the Sawamura house as a defensive system and an account of the shinobi mission against Perry's fleet. A discussion of the area of Ueno City called Shinobi-chō allows him to introduce the expression 'the ninja of Iga'. A quick canter through Sengoku history follows, in which he introduces us to *rappa* and *suppa* and finishes with the ringing statement that compared with how ninjutsu developed in other provinces, 'it was brought to perfection in the form of Iga-Kōka ninjutsu'. The next section acknowledges Iga's debt to Sun Zi's *Art of War*, and adds the now familiar claim that ninjutsu's Japanese origins were to be found somewhere within Prince Shōtoku's supposed use of shinobi and the religious tradition of the *yamabushi*, although this is as far towards mysticism that Okuse allows himself to go.

Okuse claims that the ninjutsu world was a secret organisation and states that in Iga there were many famous *ninjutsu-tsukai*, his preferred expression for its practitioners. He then introduces the notion of a ninja hierarchy consisting of lower *genin* as the 'hands and feet' for the commissioning *jōnin*. In *Mansenshūkai* the expressions had only referred to different skill levels. Okuse was the man who transformed the idea into a social structure. This may have started as a simple attempt to classify certain historical characters to whom Okuse ascribes appropriate rankings, but it was an idea that would be repeated endlessly in subsequent books. Warming to his theme, Okuse decides that there were three *jōnin* in charge of Iga during the Sengoku Period: Momochi Sandayu, Fujibayashi Nagato-no-kami and Hattori Hanzō, whose jurisdiction covered different areas of the province. In terms of their seniority the term is not unreasonable, but he then labels as *genin* ten contemporary warriors who include Otowa no Kido (the sharpshooter from *Iran ki*) and the outlaw Ishikawa Goemon. Okuse then enters the realm of complete fantasy and lists three *jōnin* from pre-Sengoku times: the magician Fujiwara Chikata, the legendary bandit chief Kumasaka Chōhan (defeated by the young Minamoto Yoshitsune), and Yoshitsune's poetic sidekick Ise Saburō Yoshimori.

141

More realistic details follow of the Momochi mansion and the site of their castle, including some features that no longer exist, but Okuse soon returns to his fanciful historical notions. His brief account of the Iga Rebellion contains so many inaccuracies that it manages to make *Iran ki* look like a work of sober history. After the war, apparently, the new *daimyō* Wakizaka Yasuharu (who was in reality the lord of Awaji Island) went on a 'ninja hunt' and *genin* like Ishikawa Goemon fled to other provinces. The fictional characters Sarutobi Sasuke and Kirigakure Saizō, says Okuse, are based on real people who served Sanada Yukimura after fleeing from Iga.

An illustrated section follows for which some very familiar objects have been photographed for the first time. The burglars' tools, saws and climbing devices are mentioned in *Mansenshūkai*. Others (including star-shaped *shūriken*) make their first appearance here in the form of photographs of real objects. The illustrations also contain what may well be the first recorded photograph of the straight-bladed ninja sword, thus proving beyond all doubt that this notorious item of ninja equipment was not a Hollywood invention, as some have claimed. The caption says that it was also made deliberately short to enable it to be used inside a house, 'thus providing no hindrance to the shinobi'.

There follows a brief summary of *Mansenshūkai*, the book that first provided Okuse's inspiration, but Okuse is not impressed by the water spider. He fully appreciates that it would not allow its wearer to skim across the surface, and adds the fascinating anecdote that at New Year some time during the Kyōhō Era (1717–35) one was tried out in the moat of Iga-Ueno Castle, a demonstration that, says Okuse, 'would have made everyone laugh', although he does not specify how it was used. The most interesting photograph in the booklet is of someone dressed as a ninja in the standard black costume that was needed 'because the ninja operated at night'. It resembles the drawing of a *ninsha* in Itō's 1937 book and must be the first time anyone was asked to pose in this way for a photograph. The booklet includes a timeline and a ninjutsu 'family tree', a version of which is still on show in the Iga-ryū Ninja Museum. There is also a long final section in which he returns to the *yamabushi* and Fujiwara Chikata to mark the end of an extraordinary presentation.

Okuse and Adachi would reuse much of *Ninjutsu* in their subsequent books. Adachi Ken'ichi published his own *Ninjutsu* in 1957, thus making it the first full-length book about ninjutsu to be produced in the post-war period. He wrote it from the point of view of a cultural historian and notes that all he previously knew of ninja in Osaka had been drawn from the fictional work of the Tachikawa Bunko. After visiting Okuse in Iga, says Adachi, he became interested in the relationship between ninjutsu as popular entertainment and ninjutsu as a martial art, and in the blurb on the inside cover he asks himself the question of why 'ninja things' (忍者物) should be flourishing on television, in films and in novels in a modern age of Intercontinental Ballistic Missiles and artificial satellites. His answer is similar to that of Okuse, in that the stories of Sarutobi Sasuke and Kirigakure Saizō appealed to the Japanese people's sense of nostalgia. Adachi's 'unique book' will shed further light on the matter, and unlike so many other authors who would follow him into print Adachi sees the historical shinobi and the fictional ones of legends and novels as equally deserving of study if a complete picture of the ninja phenomenon is to be obtained.

Adachi warns his readers that he is not a historian, and it must be said that fact and fiction tend to get mixed up along the way in his juxtaposition of references to *Iran ki* beside Fujiwara Chikata and his devils. He begins with a long account that traces the relationship between Momochi Sandayū and Ishikawa Goemon in legend and romance. We then read about Tateoka Dōjun and Otowa no Kido, along with the *onmitsu* of the Tokugawa Period. Hattori Hanzō and the occasion when the Iga-mono went on strike are also mentioned. He then moves on to Kōka and its ninja house, which leads into an account of the Kōka warriors at Fushimi and the subsequent petitions. A section entitled 'The Techniques known as Ninjutsu' draws on the military manuals, giving precedence to *Shinobi no hiden*, because he believes it to be older than *Mansenshūkai*. He confines himself to their content when he writes about ninjutsu equipment, so there are no *shūriken* in Adachi's *Ninjutsu*. He includes the same photograph of a man dressed in what he terms 'the ninja uniform' that appeared in Okuse's booklet, but notes that the idea of the black costume contradicts what was said in *Mansenshūkai*. There is a long discussion of the various manifestations of the character Sarutobi Sasuke and a fascinating chapter about ninjutsu films, including a useful list of post-war productions, which is made more interesting for the modern reader because it was written in 1957. His colleague Okuse Heishichirō makes a cameo appearance in *Ninjutsu*. Ever the showman, he was always ready to dress up as a ninja and appears in one photograph dressed in black and climbing a tree.[3]

143

Adachi would return to the subject of ninjutsu for one further book (produced with the help of two others) *Ninpō: Gendaijin wa naze ninja ni akogareruka* of 1964.[4] Some of the material and the photographs from 1957 were reused, and he also writes about the work of Okuse, who had followed Adachi's 1957 debut in 1959 with his *Ninjutsu Hiden* (Ninjutsu Secrets).[5] This was Okuse's first full-length book and the links with his *Ninjutsu* booklet are obvious, although Okuse has moved on from simply making mistakes to embracing them if the topic in question enhances the overall product. So the water spider, treated almost as a joke in *Ninjutsu*, appears on *Ninjutsu Hiden*'s opening page in the form of a posed photograph of a ninja re-enactor crossing water in the now familiar manner with one water spider on each foot. There are several more photographs in this style that resemble the movie stills that were to come only a few years later. As for the text, the first half is a history of ninja that casually includes legendary figures alongside historical ones. There is also a chapter on the links between *ninjutsu-tsukai* and the Iga Rebellion, where *Iran ki* is acknowledged.[6] Much of the rest of the book is taken from *Mansenshūkai*.

Okuse followed up *Ninjutsu Hiden* with *Ninjutsu, sono rekishi to ninja* (Ninjutsu, its history and the ninja) in 1963. In this volume Okuse returns to his totally uncritical idea of ninja history, so the Kuroda *akutō* are there along with Fujiwara Chikata and a section entitled 'Zen and Ninjutsu'. Ishikawa Goemon is mentioned in a matter-of-fact way as the man who tried to assassinate Hideyoshi and was boiled to death for his trouble.[7] The word ninja is still used very sparingly, because the protagonists in his historical accounts are still called *ninjutsu-tsukai*. In that sense Okuse's books represent a crossover between the old-style ninjutsu books and the ninja manuals that would flood the market during the 1970s. They were no less influential for that, and most of his ideas have remained unchallenged to this day, with author after author repeating Okuse's unsubstantiated conclusions as if they were revealed truths in spite of his lack of references and his willingness to stick the label of ninja or *ninjutsu-tsukai* on anyone doing anything secret or mysterious at any time in Japanese history. As noted earlier, Okuse constantly gives the impression that at the back of his mind is the idea of the ninja as a tourist attraction, and even his colleague and admirer Adachi Ken'ichi once used the words, 'ninja PR' to describe Okuse's activities.[8]

Okuse Heishichirō would succeed beyond his wildest dreams in putting Iga on the map, but his slapdash historical approach resembles

Yamaguchi's *Ninja no Seikatsu* of 1963, which I unfortunately allowed to provide most of the sources for my 1991 book.[9] Like Okuse Yamaguchi is so free with the ninja label that the scarcely disguised ninja fantasy figure is found exerting a crucial influence on numerous episodes in Japanese history. Yet Yamaguchi's influence was limited to this one book that was never translated. Okuse's ideas, his unashamed showmanship and his passionate devotion to the interests of Iga were the first ninja studies to be disseminated overseas, initially through the writings of Andy Adams and later by scores of others, so he must be regarded as the pivotal figure in the post-war development of the ninja myth. If anyone deserves the title of the inventor of the ninja it is Okuse, and his written work was only the beginning of a unique contribution to Japanese popular culture.

Ninja and ninjutsu in post-war pulp fiction

Okuse's other far-reaching contribution to the ninja myth and the prosperity of his home town was his collaboration on a novel called *Shinobi no mono* and the film that was based on it. The black and white *Shinobi no mono* would prove to be the definitive ninja movie when it was released in 1962 and placed Okuse's image of a ninja firmly in the public consciousness for all time. Both the novel and the film will be discussed in detail below, for now we note that *Shinobi no mono* was just one of many ninjutsu-related novels published at around this time that complemented Okuse's writings in the field of non-fiction ninjutsu studies. Post-war Japan had seen a revival of ninjutsu-related pulp fiction, and its ready availability as a source for potential screenplays enabled the new genre of ninja movies to appear in rapid succession once the craze had begun. These books, which included several manga, were the post-war successors to the Tachikawa Bunko series and would provide movie plots that ranged from gritty historical dramas to children's cartoon fantasies.[10]

The style and contents of these offerings were closely related to the situation that existed in Japan following her defeat in 1945. The occupation authorities blamed samurai culture for bringing about Japan's militarism in the first place, so it was hardly surprising that a ban was introduced on militaristic and ultra-national activities such as the martial arts. It was not until 1948 that the prohibition was rescinded, but the ban on martial arts was only a small part of the efforts to demilitarise Japanese society. Anything related to the samurai tradition (however it may have been perceived) was highly suspect, so a rare

145

exhibition of Japanese swords in Tokyo had to be carefully presented as a display of the artistry of sword-making.[11] Jonathan Clements gives a fascinating example of the effects of this attitude. In 1947 a new martial art called Shorinji Kempo was founded while the ban on such activity was still in place. In order to evade the prohibition (and, incidentally, to qualify for tax breaks) its founder Sō Dōshin, a clever administrator who had studied in China and practised martial arts in Japanese-occupied Manchuria, categorised Shorinji Kempo not as a martial art, but as a religion. Shorinji Kempo thus began in Japan as an expression of Buddhist philosophy. Its masters were attired (as they are today) in the robes of Buddhist priests rather than the conventional black belts of other traditions, and practised a strange sparring ritual called 'the Fists of the Arhats', a term sufficiently obscure to fool any inquisitive American onlooker. Drawing heavily on an ancient and very genuine Confucian tradition, Sō Dōshin happily explained to any investigator that his disciples' activities were actually a form of dance.[12]

The pulp fiction authors reacted to the anti-samurai stance of the Occupation in a different – though no less creative – manner, and the political and economic environment of the times helped them considerably. Their literary response to the situation was politically to the left and thus shared with the occupying authorities a suspicion of an aristocratic and authoritarian samurai class who had led Japan into a disastrous war. This Marxist interpretation of the Sengoku Period appeared to confirm that there was nothing new in upper-class tyrants called samurai dragging innocent farmers into conflicts that were not of their own making. Samurai were villains, so tales of samurai heroism were dropped in favour of a newly discovered underclass which had once suffered oppression, but had fought back against their overlords. These stories of uprisings against Oda Nobunaga proved to be as acceptable to the Occupation as Shorinji Kempo's dancing monks.

Peasant rebels, brave underdogs and lower-class martial artists who honed their skills in secret therefore became new (or at the very least newly rediscovered) heroes and struck a chord with an audience who saw parallels in their own economic position, because a nadir in both industrial and agricultural production had fed a rise in inflation and a feeling of economic gloom. The top-down attempts by the occupation authorities to control markets, coupled with the activities of black-marketeers reacting to the scarcity of goods, would have provided authors with useful contemporary parallels to their tales of remote authoritarian *daimyō* who should be opposed and predatory local gangs who should be resisted. Serious history also showed that their heroes really had existed as bandits and mercenaries, but because so much of it had been done in secret or was simply unrecorded by their betters a liberal use of imagination was fully acceptable. Popular novels about these characters therefore circumvented the ban on the promotion of the martial arts and also produced new heroes untainted by association with Japan's samurai past. The ultimate result of the authors' endeavours would be the ninja, who was of course completely untainted by anything, owing to the fact that he had only just been invented! It was an interesting situation that Clements sums up as follows:

> If the ruling class had duped the population into WW2, perhaps there had been other lies in the past. Perhaps, argued the pulp authors of the 1950s and 1960s, there was a whole forgotten underclass in Japanese history – the people who did the real work in the samurai wars, gritty, tough creatures with gadgets and athletics abilities to match this of TV spy thrillers... With the samurai aristocracy blamed for dragging Japan into WW2, ninja formed a new proletarian archetype – honest, impoverished cunning peasants, literally unseen in the historical fiction that had previously concentrated on the ruling class.[13]

The Occupation ended much sooner than anyone had anticipated, but the theme of fighting oppression continued to find resonance among the young as the 1950s wore on. Fantasy also played its part, because many of the popular novels and manga of the post-war period mirrored the earlier Tachikawa Bunko series in their love of magical ninjutsu and their choice of heroes to perform it. Sarutobi Sasuke was one of the first ninjutsu masters to be dusted off and repackaged for a new generation. A novel about him by Tomita Tsuneo was published in 1948 and may have had the reading as ninja in a book title for the first time.[14] Otherwise the first confirmed appearance of the reading dates from 1955 with another *Sarutobi Sasuke*, this time by Shibata Renzaburō, in which the characters are glossed to be read in that precise way on their first appearance.[15] The story follows the 1914 original about the celebrated ninjutsu magician.

In the field of manga an early and highly influential series was *Ninja bugeichō* by Shirato Sanpei, which began in 1957. Shirato, born in 1932, was influenced by left-wing politics and this led to his portrayal of underdogs fighting back against oppression using martial arts and weapon skills. During Shirato's childhood his father was active in the proletarian culture movement and was associated with Communist Party leaders. As he grew up Shirato experienced the horror of the war years, and these grim experiences are reflected in the nihilistic society portrayed in his works. *Ninja bugeichō* was a historically based work with vivid yet simple action drawings that captured the attention of students and intellectuals of the time. Its hero Kagemaru helps local farmers in their struggle in north-eastern Japan, but when he is introduced to the reader he comes over as less of a ninja and more of a super-samurai who uses his sword to very good effect, even though in one frame a star-shaped *shūriken* is thrown for the first time in a manga.[16] Supernatural episodes are largely avoided, and as for nomenclature the wicked lord of the castle dismisses his enemies in the first volume as '*nobushi, tōzoku* (thieves) and *shinobi no mono*', so *Ninja bugeichō* is strictly speaking a ninja manga only in its title.[17]

Ninja are also still not quite what they would become in the influential novel *Kōka Ninpōchō* 甲賀忍法帖 (the Kōka Ninpō Scrolls) by Yamada Fūtarō (1922–2001), which was first published in 1958.[18] It was not based on any earlier Tachikawa Bunko production, although it shares with works such as *Sarutobi Sasuke* a tendency towards the fantastic. *Kōka Ninpōchō* would eventually give rise to the hugely popular manga series *Basilisk*. The strange plot tells how Tokugawa Ieyasu, having unified Japan, faces the problem of which of two grandchildren will succeed to the country's throne after his son Hidetada. His bizarre solution is to make the two rival ninja clans of Kouga (sic) and Iga decide the issue by a fight between ten champions on each side. Unfortunately for Ieyasu the two clans had sworn a non-aggression pact many years before. Ieyasu's plan is to trick them into fighting each other anew without knowing of the deal he has made. There is also a Romeo and Juliet romance between two members of the opposing clans who are faced with the prospect of having to kill each other.

The book opens with a grotesque duel between two men from the rival clans who are predictably called ninja in the English translation. In the Japanese original they are not identified as such and very little of what they do is recognisable as the art of the historical shinobi. Instead it is supernatural ninjutsu of a remarkable kind. One man's weapon is a rope made from his magically treated hair; his opponent uses a sticky mucus spat from his mouth, which can trap an enemy or suffocate him. In spite of these strange features the English translator, who no doubt had marketing strategies in mind, rendered the title as *The Kouga Ninja Scrolls*, and the word ninja appears much more frequently in the English text than it does in the original version. So, for example, the single-character word *jutsu* (techniques) on page nine is translated as 'ninja techniques'.[19] The translator was probably thinking about the man in black whom he would have seen in the movies, but Yamada's book is really much more about old-style magical ninjutsu. In an appendix to the Japanese original Yamada explains why he chose to use *ninpō* 忍法 rather than ninjutsu in his original title. The *hō* in *ninpō*, he says, signifies a higher level of authority than mere techniques as in ninjutsu, extending it to laws and principles. The word was nevertheless replaced by ninja in the English version.

In spite of its fantasy features *Kōka Ninpōchō*'s reference to Iga and Kōka and its meticulous attention to historical detail might have led it to be regarded as the first 'ninja novel' were it not for the existence of another book that has a much better claim to the title. This is the afore-mentioned *Shinobi no mono* >忍びの者 by Murayama Tomoyoshi, the source of the hugely influential film of the same name, which differs from *Kōka Ninpōchō* in its anticipation of the fully human ninja who would appear in the movie. Murayama's book had only just been published when the film came out, but had been run as a serial in a Sunday newspaper between November 1960 and May 1962.[20] Adachi Ken'ichi describes it as 'a realistic romance'.[21] It is an apt description, and in an epilogue to the novel Murayama acknowledges his great debt to 'the ninjutsu researcher' Okuse Heishichirō, who had helped him so much. When Murayama was planning the book he became Okuse's guest in Iga. In Ueno City library he was introduced to *Mansenshūkai* and its illustrations, and Okuse also took him on a tour of the 'actual sites' of the historical shinobi.[22] Murayama also made use of the unpublished research notes of Adachi Ken'ichi, and the resulting *Shinobi no mono* is so dependent upon their ideas and Iga's collection of artefacts that the novel is almost Okuse and Adachi's 1956 *Ninjutsu* booklet turned into a work of historical fiction. It includes line drawings of ninja equipment, all of which are based on photographs in *Ninjutsu*. These include the *shūriken*, the nine finger signs and of course the straight-bladed ninja sword. Following Okuse again the ninja in the novel are skilled human beings, not magicians, and they are of course very firmly associated with Iga.

Even Okuse Heishichirō could not have anticipated how his generous cooperation with an enthusiastic popular novelist would spread his own Iga-centric ideas of a non-magical type of ninja to a wider audience than anyone would ever have imagined. That happy result came about when *Shinobi no mono* became a film, and in the hands of the movie's director the benchmark ninja sprang from the written page and came to life in thrilling action. Things were never quite the same again, and the change that took place is summed up by the covers of two different books involving material by Fujita Seikō. The cover picture of *Ninjutsu Hōten* of 1955 features a romantic-looking youth dressed in beautiful attire. *Hiden ninjutsu no hon* by Akira Nakao and supervised by Fujita dates from only eight years later, but the figure on the cover is a fully formed black-clad ninja escaping acrobatically from his pursuers.[23] It is now 1963, and a revolution has occurred.

Chapter 12

Enter The Ninja

The medium of motion pictures gave the newly created ninja the opportunity to leave the printed page and find expression through the universal language of live action. What is remarkable is the short space of time in which it all happened, because a creative explosion occurred between late 1962 and early 1964 that defined the ninja forever both in terms of the concept, and also in the extent of its reach.[1] It is nevertheless somewhat difficult to define a ninja movie, because if we go solely by the inclusion of a reading as ninja in its main title no film had this until *Jūshichinin no ninja* 十七人の忍者 (Seventeen Ninja) in 1963. To confuse matters further, when the ninja craze began to take hold a massive relabelling exercise took place within the film world's back catalogues. For example, the mundane samurai movie *Yaro no Maden* of 1956 now appears in modern filmographies as *The Ninja's Weapon*, even though there are no ninja in it.

In 1962 the pivotal *Shinobi no mono* stayed with the old reading of shinobi in its title, even though the word ninja is heard throughout in the spoken dialogue. This reinforces the point made earlier that the reading is less important than the word's significance, but *Shinobi no mono* definitely broke new ground, as becomes apparent when a comparison is made with the films that anticipated the revolution it brought along. In several post-war movies elements were included which we now almost automatically associate with ninja, such as a masked hero dressed in black and a wide range of activities called ninjutsu. However, just as in the earlier books and manga, the ideas that lay behind these elements were somewhat different from what they would eventually evolve into. The characters in *Shinobi no mono* would be much less fantastic than the mucus-spitting ninja in Yamada's novels,

151

so if anything the film's great achievement was to tone down or even remove the pulp-fiction fantasy to make the ninja more human. Other directors would of course put the fantasy back in during later productions, but by then the black-clad human ninja shot in appropriately harsh and shadowy black and white established a conceptual norm that he retains to this day.

Ninjutsu had kept its magical meaning as late as 1957 in the movie *Ninjutsu gozen-jiai* 忍術御前試合, in which the hero is the legendary outlaw Ishikawa Goemon under his youthful name of Torawakamaru. His particular skills at ninjutsu make him a sorcerer and a shape-shifter rather than a secret agent, and in a clever and very entertaining sequence he demonstrates his prowess in front of Toyotomi Hideyoshi. Torawakamaru first behaves like a magical drill sergeant to make a squad of samurai freeze, run in circles and generally do his bidding. When they attack him *en masse* with their swords he shape-shifts into a stone lantern and then reappears in human form on the roof. Torawakamaru finishes the performance by making everyone think that the adjacent keep of Osaka Castle is on fire. The one hint that the film provides towards the later development of the ninja movie is to introduce Iga and Kōka as rival clans.[2]

The 'ninjutsu as magic' theme was very important in providing building blocks for the invention of the modern ninja, but one thing it did not do was to add anything to the future ninja's visual image as a man dressed from head to foot in black. The enveloping costume starts to appear in films from the 1950s onwards where subterfuge is involved. When we examine them nowadays the image of the ninja seems so obvious, but just as in Tokugawa Period illustrated books it is the image of a man of secrecy that is being presented, not a specialised shinobi no mono. One remarkable example of an anticipatory film along these lines is *Sarutobi Sasuke Senjōgadake no himatsuri* 猿飛佐助千丈ケ嶽の火祭 of 1950, based on the hero from the Tachikawa Bunko. Many of the scenes herald the human ninja scenario that would appear over a decade later, and some of the fight scenes are tending in that direction, but it is in the costumes that we see the biggest difference from the Jiraiya movies. The hoods and tight leggings of the heroes make their wearers ninja in all but name and Keith Rainville regards the film as an eye-opener because so many ninja elements are ready for exploitation. He does, however, draw a contrast between the film's combat scenes, which still depended greatly on the camera trickery of the Jiraiya mode, and the martial arts-based ninjutsu that would come in later.[3]

Sarutobi Sasuke Senjōgadake no himatsuri was not the only 1950s film to anticipate the fully rounded ninja in its choice of costume. *Kaiketsu Kurozukin* 怪傑黒頭巾 ('the wonder man in the black hood') is a phrase that appears in the titles of two films released in1955. They are set in 1867 and the mysterious black-hooded man whose identity no one knows looks exactly like a ninja, although in all other aspects he is more reminiscent of the Lone Ranger, particularly due to the fact that he packs pistols. The character was still being portrayed into the 1990s and stills from that time show what appears to be a gunslinging ninja riding a white horse, which suggests that the inspiration for the character came from Hollywood westerns rather than Japanese legends. Other hooded men include the bizarre superhero in the television series *Gekkō Kamen* that ran from 1958 to 1959. He was masked and dressed all in white with a turban and sunglasses. In place of a horse he rode a motorcycle.

The separate themes of the hooded warrior and the ninjutsu superhero would soon come together to produce the live-action human ninja movie. Looking back from a position of familiarity the amalgam seems so obvious that one wonders why it took until 1962 for it to be realised, and it is interesting to note that the two themes had already blended before this date within the medium of films for children. The most important example was *Shōnen Sarutobi Sasuke* 少年猿飛佐助 (the boy Sarutobi Sasuke), an animated feature film that appeared in December 1959. It was released in North America on 22 June 1961 under the title of *Magic Boy* and is a reminder of how important children's material would be in the creation of the ninja image. *Magic Boy* was the first ninjutsu movie to be seen in the USA, but this Disney-like children's drama was a far cry from the ninja movies that would follow only a year later. Indeed, any ninja connection seems to have been deliberately played down. MGM's publicity apparently stated incorrectly that the film's original Japanese title was 'The Adventures of the Little Samurai' because they preferred to associate their subject with the noble heroic image of the samurai than the supposedly sinister image of the ninja.[4]

The ninja aspects of *Magic Boy* may have been played down in America, but it was not long before Hollywood ate its words, because in late 1962 the crucial elements of the black costume, the ninja as martial artist and the association with Iga and Kōka came together in the movie *Shinobi no mono* to create the ninja as we know him today. Like Murayama's original novel, *Shinobi no mono* was a historical drama based on real events. It was released on 1 December 1962 and proved to be so popular that no fewer than seven sequels were produced, the last appearing in December 1966.[5]

At about the same time a very influential television series also began. It was quite different in concept and profoundly different in its target audience, but played an equally important role in the creation of the image of the ninja and everything that derived from it. The series was called *Onmitsu Kenshi* 隠密剣士 (the swordsman spy) and was presented as a family show suitable for children. It was produced by Nishimura Shunichi, whose earlier credits included the masked motorcyclist in *Gekkō Kamen*.[6] *Onmitsu Kenshi* began on 7 October 1962 and was screened at 7pm on Sunday evenings. If we include its follow-up series *Shin Onmitsu Kenshi* (new swordsman spy) the whole production ran until 26 December 1965 without a break, except for a brief hiatus caused by the Tokyo Olympics of 1964. It was also dubbed into English and enjoyed a very successful run in Australia as *The Samurai*. As both *Shinobi no mono* and *Onmitsu Kenshi* were to become so influential, a trivial debate may be had about which actually came first. *Onmitsu Kenshi* was technically the forerunner because it appeared on the television screen on 7 October 1962, but its first series of thirteen weekly episodes involved no ninja at all. It was only with the beginning of the second series (called *Ninpō Kōka-shū* 忍法甲賀衆 and rendered into English as 'Kōka Ninjas') on 6 January 1963 that ninja made their first appearance, by which time the ninja movie *Shinobi no mono* was already well-established in the public consciousness.

Shinobi no mono and *Onmitsu Kenshi* have two important things in common. The first is that the word ninja is used throughout in the spoken dialogue yet appears in neither of the main titles. This may have been of little consequence for Japanese audiences but, almost needless to say, the word ninja was added to the English-language versions, although its deployment was still somewhat limited. The subtitled *Shinobi no mono* became *Ninja Band of Assassins*, but when the dubbed *Onmitsu Kenshi* was released in Australia in 1965 the series was not called *Ninja*, as one might have expected, but was given instead the unimaginative title of *The Samurai*, while the Japanese feature film of 1964 based on the series is now available on DVD with English subtitles under the literal yet uninspiring translation of *The Detective Fencer*. The second thing they have in common is their presentation of the ninja tradition as an ancient revealed truth, even though the stories were actively and consciously constructing it as their plots developed, just as Kikuoka had done in 1679 with *Iran ki*.

The 'ninja-free' first series of *Onmitsu Kenshi* is set in Hokkaido in 1792. The action is reminiscent of a Western, with the Japanese settlers playing the cowboys and the native Ainu the indians, and the only ninja-like element comes from the use of spying as a theme, because the samurai hero Akikusa Shintarō is an official government *onmitsu*. In a plot-line with a sound historical basis he has come from Edo to gather information about the *daimyō* of Matsumae (the only domain on Hokkaido at the time) on behalf of the Shogun. The fighting involves swordplay with no super-human dimensions at all, but by the time the first programme of the second series begins everything has changed, which makes one think that the director must suddenly have realised what was missing. A sword-wielding *onmitsu*, regardless of the fact that he represented historical reality, was clearly not enough. It was too late to transform Shintarō into a ninja, so ninja were introduced into the new series as the bad guys, and as the lights fade during the opening credits a ninjutsu scroll opens while a voice intones information about ninja and their code of behaviour. Various ninja weapons then appear as stills and are soon deployed in exciting action sequences that pit the noble official *onmitsu* against a mysterious secret society of ninja. From that moment on *Onmitsu Kenshi* became the most influential ninja television series of all time, although its claim to have been the first ninja action drama had been nullified by the release of *Shinobi no mono* a month earlier, while *Onmitsu Kenshi* was still conceptually stuck on Hokkaido.

Having pipped *Onmitsu Kenshi* to the post, *Shinobi no mono* had set the ninja tone from the word go, but the image it presented was different from anything that had gone before. No longer were the heroes ninjutsu magicians who threw bolts of lightning and turned themselves into toads; nor even were they Yamada's strange superhuman warriors who used their hair as weapons. Instead characters in black hoods, whose predecessors had worn similar attire merely to conceal their identities, became warriors called ninja who hailed from Iga Province and exploded into furious action in ways that were almost completely subject to human limitations. This image would be firmly associated with the ninja for evermore, and because a feature film of the 1960s possessed a wider reach than any television series, *Shinobi no mono*'s pivotal role in the movies and beyond can hardly be overestimated.

Many of the ninja weapons and tools that the film was to publicise so successfully had been illustrated in Okuse's *Ninjutsu*, so it was in everyone's interests that he should be enlisted in a personal capacity to help with the movie's production. Okuse's belief that Iga was the centre of the ninja tradition was a vital part of his vision, and the film would set in stone his notion of an indelible link between ninja and his birthplace. Inside the Iga-ryū Ninja Museum that he founded is a display case that commemorates his writings and his work on *Shinobi no mono*. There are movie posters for it and a photograph of Okuse with its star Ishikawa Raizō standing beside Iga-Ueno Castle.[7] Okuse's imaginative research therefore gave the film its overall look, its Iga-centred environment and much of its material culture, all of which provided a considerable boost for the Iga tourist industry.

The screenplay for *Shinobi no mono* was of course based on Murayama's novel, but the other crucial influence in creating *Shinobi no mono* was one for which both Okuse's and Murayama's professional input was necessarily limited. This was how the ninja would move and fight, and the producers wanted something different from conventional samurai movies. Ninja combat (whatever that was) needed to employ Okuse's full range of ninja devices and be conducted at a much faster pace than the long stand-off and quick denouement of the classic samurai movie duels. To achieve this goal the film company engaged the services of the martial artist Takamatsu Toshitsugu (1889–1972) and his pupil Hatsumi Masaaki as advisers for the action scenes.

The role of adviser for a film can be challenging, rewarding and often frustrating, as I learned in 2010 when I acted as adviser for the movie *47 Ronin*. That film was a fantasy based on historical reality, and I accepted fully that fantasy would trump history if my advice conflicted with the drama that was intended. The role of the historical advisers for the 1960s ninja movies appears to have been very different, because they did not merely advise, but became partners in a creative act. Takamatsu acted as the overall adviser, while young Hatsumi concentrated on the action scenes, using his experience of a wide range of classical combat situations and applying them to the newly created ninja. The carefully choreographed set-pieces were very well done and revealed Hatsumi's understanding of what real combat would have been like if it had ever been performed by Okuse's archetypal ninja. In *Shinobi no mono* his extensive knowledge of *bujutsu* was combined for the first time with the smoke bombs and collapsible ladders pictured in *Mansenshūkai* and displayed in the Iga museum collection. In Hatsumi's hands the hooded heroes from the 1950s were recreated as ninja and allowed to demonstrate just enough mystical ninjutsu lore to add a suitable level of esotericism to their superlative swordsmanship, while staying sensibly within the technical limitations of the weapons and authentic human capabilities. Hatsumi's practical contribution to the ninja myth about how the ninja would fight neatly complemented the very different type of work done by Okuse Heishichirō and was almost as influential in creating the ninja as we know him today. It was a consummate exercise in standard-setting, and Hatsumi went on to become one of the most influential people to develop the practical side of ninjutsu. He would accept other advisory commissions including the James Bond film *You Only Live Twice* and once even played a ninja in a film. His later creation of the Bujinkan organisation and its worldwide training programmes would lead to ninjutsu being widely understood as a martial art in its own right. All this provided a further layer of normalisation to the film's quasi-historical definition of the ninja's social position and his links to Iga. That was the film's other great achievement: to place its ninja creations firmly in the environment that Okuse had created for them. *Shinobi no mono* therefore influenced not only how ninja would be represented on screen, but also how they would be understood and portrayed far into the future, from souvenir key rings to expensive ninjutsu training courses.

Like so many ninja films that would follow it, the historical events around which the plot of *Shinobi no mono* is set are the Iga Rebellions of 1579–81. The film is aimed at adults rather than children, as is readily apparent from the inclusion of scenes of people being crucified or having their ears cut off. There is also a great deal of political intrigue that is explained through dialogue, so it is by no means one long set-piece ninja combat. The opening titles tell us that it is the year 1573 and Oda Nobunaga, who intends to conquer all of Japan, has been fighting the Azai and Asakura in northern Ōmi. The scene behind the titles is a smouldering battlefield where we encounter two wounded samurai, one of whom obligingly does a back somersault to show us that he is a ninja. The shot then fades to a map indicating that we are in Iga Province, where two ninja clans are planning the death of Nobunaga, a monster of a man who is played with sadistic relish by the actor Wakayama Tomisaburō.

As in Murayama's original novel, the central heroic character is the outlaw Ishikawa Goemon. He becomes involved with the rival ninja clans who are called Momochi and Fujibayashi, the names having been borrowed from historical Iga families. This is the first of the film's defining moments because their clan members are neither *kokujin* nor *jizamurai*, and they do not fight in the Iga-shū. Instead we have left the authentic political structure of Iga far behind and are confronted by two rival ninja families who appear to be neither samurai nor farmers, but an élite, if socially humble, hereditary cult. This encompasses once again the theme of the ninja as heroic underdogs fighting oppression by a wicked overlord, just as in the words of Shibata Renzaburō and the drawings of Shirato Sanpei. With that point firmly established, the historical context in which the action unfolds is broadly correct and the film finishes with the second Iga Rebellion of 1581. Along the way the movie paints Okuse's picture of ninja life, which would become the conceptual norm. It is stated very early on by the Momochi leader that ninja sell their services to other *daimyō*. Ninja are therefore firmly defined as mercenaries, and among their numerous martial accomplishments they make their own gunpowder and train fiercely. The ninja also perform the nine mystical hand movements indicative of esoteric ninjutsu and every attempt at castle entry has them dressed from head to foot in black. One swims a castle moat using a bamboo tube as a snorkel. *Shinobi no mono*'s ninja also throw star-shaped *shūriken* at each other within twenty minutes of the film's commencement, so within the space of one hour the three great stereotypes of the ninja: the black costume, the finger signs and the *shūriken* have made their first appearances together in a live-action drama.

In *Shinobi no mono* the hated Oda Nobunaga survives three attempts at assassination. In the first scenario his pet cat unfortunately receives the dart-like *shūriken* intended for him, while the others are ninja classics that would become plot stereotypes for future movies. In the second attempt poison is dripped down a thread into his mouth, and in the third the assassin lurks under the floorboards of Nobunaga's bedroom waiting to strike upwards with his sword. He is, however, betrayed by the ringing of a bell set off by a trip wire, and in a dramatic finish one of Nobunaga's bodyguards jabs his sword through the *tatami* mat. *Shinobi no mono* closes with Nobunaga attacking the ninja headquarters in Iga in a tremendous battle scene spoiled only by the anachronistic use of projectiles that behave like high-explosive shells. A nice touch is the introduction of caltrops, now stock-in-trade of any ninja museum's display, which bring Nobunaga's troops to a halt to be hit by a hail of arrows.

As *Shinobi no mono*'s image of a ninja is now so familiar, it is a fascinating exercise to compare the approach taken by its television rival when the second series of *Onmitsu Kenshi* introduced its own take on ninja six weeks later in January 1963. *Onmitsu Kenshi: Ninpō Kōka-shū* was designed for a much younger audience than *Shinobi no mono*. Its influence was therefore different, but just as important because so much of ninja culture has been aimed at children, and one big difference brought about by the choice of target audience concerns the time periods in which they are set. Gone are the horrors of war in the Sengoku Period. Instead *Ninpō Kōka-shū* takes place in 1792. There are no bloody battles to be fought and when the heroes are not outwitting the secret society of the Kōka ninja they hunt for stolen gold.

The story begins with the government agent hero Shintarō back in Edo after his mission on Hokkaido, and the opening sequences of the ninja scroll and a sombre voice-over have already suggested to the viewer that it will not be long before he comes upon a ninja. This is so, because a ninja from Kōka kills a member of the Shogun's Oniwaban, but even though the Kōka ninja is dressed in black like the Iga ninja in *Shinobi no mono*, the gap of 200 years means that they are conceptually different. The *Shinobi no mono* ninja had been members of rival clans. The newer variety are presented as members of 'the secret society of the Kōka ninja' who must be confronted by Shintarō and the good guys from Iga who are the Shogun's loyal followers. The first combat begins with the throwing of *shūriken* and proceeds to a hooked rope with which the Kōka ninja pulls his victim forwards to use his sword against him. When challenged by Shintarō the ninja does a series of back flips, releases a smoke bomb and then flies slowly backwards up on to the roof. With that one superhuman movement, completely absent from *Shinobi no mono*, the human ninja has become a wizard once again, and just as the ninja of *Shinobi no mono* were defined right from the start as very human mercenary members of ninja clans, so *Ninpō Kōka-shū* redefines them as supreme martial artists who demonstrate this little extra touch of magic.

Throughout *Onmitsu Kenshi* the very human Shintarō has to fight against wicked superhuman ninja and, it has to be said, he fights them very well. In one scene he spots a ninja in the ceiling above him by seeing his reflection in a cup of tea, but even Shintarō is bemused temporarily by a ninja spinning at great speed, a technique identified in the English subtitles as 'ninja dazzle'. By Episode Two the ninja villain is able to change the appearance of his face using magic. The fantastic aspects of the ninja then grow with every passing series, and even though they never reach the level of Jiraiya's sorcery, some underwater dragon-ships eventually make an appearance. Nevertheless, the word ninjutsu now has an entirely different meaning from Jiraiya's magic, because no matter how fantastic a gadget may be it represents ninjutsu as the art of the ninja, an amalgam produced by combining established martial arts with supposed ninja weapons. *Onmitsu Kenshi* therefore brings *Shinobi no mono*'s ninja out of the Sengoku darkness and adds just enough fantasy to appeal to a wider audience.

Onmitsu Kenshi grew increasingly popular week after week, so the creators of *Shinobi no mono* acted quickly to mark their territory in the world of 'adult' ninja films. *Zoku Shinobi no mono* (*Shinobi no mono continued*) was released on 10 August 1963 and a third film, *Shin Shinobi no mono* (new *Shinobi no mono*) followed rapidly on 28 December of the same year. The second story follows the first seamlessly with the conclusion of the Iga Rebellion. The attempt to shoot Nobunaga at the Aekuni Shrine (its name is mistranslated in the subtitles) is very well done, but Ishikawa Goemon has sworn to kill Oda Nobunaga all by himself, so he has to be squeezed into the historical circumstances of Nobunaga's death at the Honnōji in 1582. Akechi Mitsuhide's surprise attack is brilliantly performed with Goemon administering an unhistorical *coup de grâce*. Towards the end of the film Goemon also carries out an unsuccessful assassination attempt on Hideyoshi, who is surprisingly found to be sleeping Western-style in a brass bedstead. Goemon is captured, but whereas the historical Ishikawa Goemon was executed by being boiled to death, the ninja Goemon somehow avoids the bubbling cauldron to survive until the third film. When Goemon is finally dispensed with by the fourth sequel *Shinobi no mono* needed a new hero, and one was found in the person of Kirigakure Saizō, another fictitious character from the Tachikawa Bunko series whose familiarity had made him into almost a historical figure.

Shinobi no mono and *Onmitsu Kenshi* are the quintessence of the ninja myth. Within them is encompassed the entire conceptual range of the benchmark ninja as he is now understood, from harsh reality to cheerful fantasy, and from ninjutsu teaching qualifications to the production of ninja toilet paper. Any subsequent manifestations of the ninja have followed the broad directions to which Okuse and Hatsumi pointed so unmistakably, as was shown when other film companies rushed to share in the success. At the end of 1963 a film called *Fukurō no shiro* 梟 の城 (Castle of Owls), based on the novel of the same name, was released to show ninja in glorious technicolor for the first time. It begins with the bloody conclusion of the Iga Rebellion in 1581, and its opening sequences are highly reminiscent of *Shinobi no mono*, as is the portrayal of the ninja clans of Iga as underdogs. A major theme is the rivalry between two men from Iga, one of whom has renounced his former life and become a samurai, thereby stressing the difference between the two traditions as far as the movie industry was concerned. There is also love interest with a female ninja, a remarkable *shūriken* duel inside a forest of green bamboo and some brilliant short set-pieces including a ninja secretly entering a castle.

Other ninja films tried different tacks. *Sengoku Yaro* 戦国野郎 depicted ninja as a gang of gun-runners, while in *Iga no Kagemaru* 伊賀の影丸 (Kagemaru of Iga) a manga became a ninja movie, with once again the theme of Iga–Kōka rivalry. Like Shirato Sanpei, its original creator Yokoyama Mitsuteru was a master of ninjutsu-based manga and the film became a hit live-action film which stayed faithful to its roots. *Jūshichinin no Ninja* 十七人の忍者 (Seventeen Ninja) then put ninja into the title of a film for the first time. It is a remarkable work in which the ninja are not peasant underdogs, but the loyal Iga-mono of Edo Castle in 1631. The second Tokugawa Shogun Hidetada is dying and there are fears that his younger son Tadanaga is planning a coup to become Shogun in place of his elder brother Iemitsu. Seventeen ninja are sent to infiltrate Tadanaga's castle of Sunpu and bring back the conspirators' signed pledge that will be vital evidence for convicting them. Sunpu's security is under the supervision of a rival ninja from Negoro, and the contest between him and the seventeen good guys is as much a battle of wits as a fight with weapons.

During 1963 the ninja craze entered the public consciousness in a different way when a sixteen year-old youth climbed into the old Imperial Palace in Kyoto dressed as a ninja.[8] The incident cannot have done any harm to the brand, and in 1964 *Daisan no ninja* 第三お忍者 (The Third Ninja) added more on-screen action, to be followed by the humourless and violent *Ninja gari* 忍者狩り(The Ninja Hunt). Also in 1964 Shirato Sanpei wrote a television series for children called *Shōnen Ninja Kaze no Fujimaru* 少年忍者風のフジ丸 (Fujimaru, the boy ninja of the wind). The manga bore the original title of *Kaze no Ishimaru*, but was renamed *Kaze no Fujimaru* in order to associate it with its sponsor, Fujisawa Pharmaceuticals, and one can only imagine the effect the decision had on the committed left-wing author. The story begins when Fujimaru is a little baby and is kidnapped by an eagle. A samurai recovers the baby and takes him as his disciple, who can control the wind through ninjutsu.

The series was a well-balanced mix of genuine historical weaponry and credible martial arts with superhero magic, but even though this was a cartoon film short vignettes were added to the end of some episodes featuring Hatsumi being interviewed about ninja topics. In one he throws *shūriken* and in another he is asked about ninja scrolls and the belief that they contained spells to make one invisible. Hatsumi unwraps such a scroll and explains with a smile that the scroll is about forecasting the weather and that the ninja would choose a foggy day so as not to be seen (hence the invisibility!) although he rather spoils the joke and his own claim to being an authority on ninjutsu by stating that the scroll is about the ninja Kirigakure Saizō, who was of course a fictional character.[9]

By this time the growing ninja craze was beginning to attract some criticism from purists who felt that ninjutsu was being tainted by commercialism. Among them was Fujita Seikō, although there may be a touch of hypocrisy in his comments. He had already been responsible for inventing the ninja parachute and had willingly assisted Okuse with the exhibition in Iga, so even though Fujita never lowered himself to serving as a movie adviser, he was not entirely free from the influence of popular culture. Nevertheless, in an interview published both in Japan and abroad in 1964 he openly rebuked anyone who appeared to be exploiting the ninja tradition for money. The article appeared in *Newsweek*, where the anonymous journalist writes, 'Fujita, who claims he is the last of the Ninja and that the secrets of the craft "will die with me" deplores the current commercialisation of Ninja in Japan'.[10]

Yet the ninja boom could not be stopped, even by a legendary figure like Fujita, and in 1965 very high standards of ninja combat were set by the film I*bun Sarutobi Sasuke* 異聞猿飛佐助. It is available with English subtitles as *Samurai Spy* and is a visual masterpiece with breathtaking cinematography that makes up for the leaden dialogue and the rather odd character who is a ninja dressed all in white. Also in 1965 *Onmitsu Kenshi* began conquering Australia as *The Samurai*, with exciting ninja combat scenes and some subtle differences from the original Japanese version. Its distributors clearly realised that the absence of ninja from the original first series of thirteen weekly episodes could prove to be the kiss of death, so *The Samurai* begins at Series Two. It also has different theme music from the original. Every episode of the Japanese *Onmitsu Kenshi* began with a patriotic-sounding marching tune sung by a children's choir to a western-style backing track. In its search for a more appropriate ambience for a foreign audience *The Samurai* replaced it with a 'plinky-plonky' oriental musical cliché of the type invariably chosen for any documentary on Japanese martial arts.

162

Their finest hour

Meanwhile, *Onmitsu Kenshi*'s big screen rival *Shinobi no mono* rumbled on, and within two years (if anecdotal information is to be believed) its influence would be spread wider than ever before when images and scenes from it were reused in the 1967 James Bond film *You Only Live Twice*.[11] This was the film that effectively introduced ninja to Western audiences. It was based on the original novel by Ian Fleming, but the screenplay diverged considerably from it in many aspects, including the action sequences involving ninja. As for its background, Fleming once spent three days in Japan where one of his hosts was a martial artist, as he describes in his travelogue *Thrilling Cities*.[12] Ninja may have been mentioned to him then, but Fleming makes no reference to them until ninja appear in the novel *You Only Live Twice*. The following passage paints a vivid picture:

> there came a whistle from above them on the ramparts and at once ten men broke cover from the forest to their left. They were dressed from head to foot in some black material, and only their eyes showed through slits in the black hoods. They ran down to the edge of the moat, donned oval battens of what must have been some light wood such as balsa, and skimmed across the water with a kind of skiing motion until they reached the bottom of the giant black wall. There they discarded their battens, took lengths of rope and a handful of small iron pitons out of pockets in their black robes and proceeded to almost run up the walls like fast black spiders.[13]

The black costume worn by these characters was shortly to be revealed to the Western world in the first visual adaptation of *You Only Live Twice*. This was, however, not the 1967 film, but a comic strip version of the novel serialised in the *Daily Express*. Between 1965 and 1966 the artist John McLusky produced what were probably the first drawings of ninja that anyone outside Japan had ever seen. McLusky's take on the ninja suit differs somewhat from the benchmark figure, but his phantom-like characters are a significant footnote in the history of ninja media, and as Keith Rainville points out in his commemorative review of *You Only Live Twice*, McLusky's figures were also a good deal more like ninja than the strange characters in grey who would appear in the subsequent film.[14]

The screenplay for the movie of *You Only Live Twice* was written by the famous children's author Roald Dahl. He visited Japan along with the production team and was present on set, and according to the liner notes for the 2007 DVD version of *Shinobi no mono* Dahl saw it sometime during his visit.[15] This is highly likely to be true, simply because certain parts of the plot of *You Only Live Twice* are very similar to scenes in *Shinobi no mono*, including the fact that both villains (Oda Nobunaga and Ernst Blofeld) have a fondness for cats, but what is surprising about the Bond film is how clumsily the ninja theme is handled. This happened in spite of the services of Hatsumi Masaaki, whom Albert Broccoli had asked to be the film's adviser for the action scenes. Hatsumi, who has a small walk-on part in the movie, was an inspired choice, but one suspects that much of the advice he gave was ignored because of the demands of Hollywood.

The first scene involving ninja in *You Only Live Twice* shows them training in the courtyard of Himeji Castle and is reminiscent of a similar scene in *Shinobi no mono*. Himeji, which the viewer sees first in an aerial view, is presented most alluringly as a 'training school for ninjas' (sic), yet it is a place where no one dresses in black and the only ninja weapons visible beside the conventional implements and actions of kendō and karate are *shūriken*. The throwing of them at targets (one certainly misses) apparently caused concern that they might damage the walls of Himeji, until it was pointed out that the set designer had built a false wall in front of the real one.[16] The ninja leader in the film is also so bad at ninjutsu that he allows two outsiders to infiltrate his castle and make attempts on James Bond's life, one of which is by means of poison dripped down a thread. Exactly the same scene appeared in *Shinobi no mono*.[17] The ninja action only really gets going in the famous volcano scene, even though the black-clad ninja envisioned in Fleming's original novel and drawn in the *Daily Express* have been replaced by what look like commandos dressed in grey who abseil down into the crater with samurai swords across their backs. The swords are used during the fighting in addition to guns, and there are two memorable sequences with thrown *shūriken*. One kills a guard and the other makes Blofeld drop his revolver. This must have fixed the weapons indelibly in many a mind.

For most foreign audiences the ninja, the *shūriken* and even the conventional Japanese martial arts sequences would have been the first time they had ever seen the like. *You Only Live Twice* was therefore truly groundbreaking in presenting these concepts outside Japan, even though the ninja image had been considerably modified for Western tastes. The result was that by 1967 it was still only the fortunate inhabitants of Australia who could enjoy classic ninja action by viewing the dubbed *Onmitsu Kenshi* as *The Samurai*. Everyone else, whose appetite was whetted by *You Only Live Twice*, would have to wait a few more years to see 'real ninja', and perhaps only the readers of the martial arts magazine *Black Belt* knew what they were missing,[18] because between December 1966 and February 1967 *Black Belt* published an innovative series of three articles about ninja by Andy Adams.[19] The January 1967 edition provided my own introduction to the world of the ninja, and I was stunned by the exciting ninja movie stills, the photographs of Hatsumi in action and the pictures from museums of apparently authentic ninja weapons. Much of the material for the features was drawn from the writings of Okuse Heishichirō, as Adams later explained.[20] These articles pre-dated the release of *You Only Live Twice* by only five months. Having reread them fifty years later I now appreciate fully what might have been done with the James Bond film if Ian Fleming's original writings had been properly exploited along the lines of Okuse's vision and Hatsumi's action.

In a follow-up article published at the time of the film's release Adams gives an interesting perspective on the situation. He sees the ninja and their 'ninjitsu' as just another James Bond gimmick to add to machine guns in cars and dart-firing cigarettes, and compares the much greater attention that was given to training Sean Connery in the martial arts. World-famous instructors were falling over themselves for the honour and Donn Draeger was the first to be hired. The famous Oyama Masatatsu ('Mas Oyama') also claimed to have trained Connery in secret and certainly presented him with an honorary third dan in public. Not to be outdone, a certain Nakashima Shohitsu, who took over 007's training from Draeger, was so impressed by Sean Connery's progress that he awarded him a real second dan, not an honorary one. It is hardly surprising that ninja did not get much of a look in.[21]

Nevertheless, *You Only Live Twice*, which was released on 12 June 1967, was truly ahead of its time in its introduction of ninja to the West and its depiction of the Japanese martial arts. The film may have been something of a missed opportunity, but it unquestionably spread the name of ninja and their ninjutsu around the world to a new generation born since World War II who would not immediately associate Japanese edged weapons with wartime atrocities. Instead Japan came across as an optimistic and futuristic society with its roots in an exotic and exciting martial past. The ninja in the film may have revealed only a little of their real selves, but the seed had been successfully sown. In Jonathan Clements' well-chosen words, 'it was their finest hour'.[22]

165

Chapter 13

A Star is Born

It is the primary thesis of this book that the development of the ninja myth has involved a constant reinvention of tradition over the course of several centuries. This argument will be developed in the present chapter by a case study of the ninja weapon *par excellence*, the device called a *shūriken* 手裏剣 in Japanese and popularly known in the West as a 'ninja star'. The image of the sharp-bladed *shūriken* spinning towards an enemy is as indelibly associated with the ninja as the samurai is with his proudly and firmly held sword, and even if the *shūriken* has never been called 'the soul of the ninja', its design has made it into the ninja's perfect defensive companion. This is because it is a small hand-held weapon that can easily be concealed when carrying out a mission, so it fits perfectly with the ninja's cloak and dagger image. In movies a hail of *shūriken* is quite often used to cover a ninja's withdrawal as he attempts to return alive with information. Otherwise, if a ninja is challenged by a swordsman a single *shūriken* whizzes expertly from the ninja's hand, making the assailant drop his weapon when the spinning blade hits his wrist.

The historical authenticity of the *shūriken* is now accepted so uncritically that even the publishers of a recent academic study of the Iga-mono of Edo Castle felt obliged to include some *shūriken* on the cover picture of an otherwise desk-bound official to indicate the man's links to the ninja of legend.[1] The throwing of ninja stars has also long been a staple of the fairground side of ninja theme parks, and target practice with *shūriken* is now advertised as a key feature of several guided tours of the Mount Fuji area. Tourist buses call in at the Oshino Ninja village so that slightly baffled foreign visitors can use the *shūriken* range after visiting the ninja fun house.

The word *shūriken* is found in several historical texts but, as with so many other aspects of the ninja myth, the meaning then differed considerably from that of today. One of the earliest uses of the expression occurs in *Gunpo Jiyoshū* of 1653, in which a *shiriken* しりけん (i.e. *shūriken*) is a torch made from split wood and fitted into a metal base with a spike. It would be thrown down by a sentry from the walls of a castle to provide illumination during a night attack. The spike would ensure it stuck in the ground, and the name given to it indicates no more than that it was a device thrown by hand.[2] Otherwise a *shūriken* was a weapon, and these were often improvised devices. Samurai did not spend much time throwing things, but even their short swords could be flung at assailants when all else failed. There were no real Japanese equivalents of the javelin apart from some specialised (though rarely used) throwing arrows called *uchine* 打根, but there did exist one historical weapon that shared the name of *shūriken* and was destined to enter the ninja armoury along with its star-shaped equivalent. This was the *bō-* (棒 i.e. 'rod- or bar-shaped') *shūriken*, a straight, heavy steel spike for which a translation as 'dart' is perfectly adequate. The expression 'throwing a *shūriken* by hand' appears in a document of the Nabeshima family from 1695 and probably refers to the darts.[3] There is also a drawing of straight steel darts being thrown against a target of a human face in an illustrated book called *Shashin Gakuhitsu* by the artist Maki Bokusen (a.k.a. Gekkōtei, 1775–1824).[4] By 1900 the word was so well known that in the novel *Hisho no tomo* by Emi Sui'in one of the characters throws a *shūriken* of some sort.[5]

Straight darts and spiked torches called *shūriken* are therefore authentic historical implements, but no spinning ninja stars appear in any of the shinobi no jutsu manuals. To the ninja skeptic that means that they did not exist and are completely modern inventions. To the ninja enthusiast their absence indicates that they must have been top-secret weapons. The reality of the situation has nothing to do with either of these extremes, and the true answer explaining their absence sheds a great deal of light upon the notion of the ninja as an invented tradition.

Ninja stars appeared quite late on the ninja scene. Seemingly unknown to Itō Gingetsu, they make their first appearance in a modern Japanese text in Fujita Seikō's *Ninjutsu Hiroku* of 1936, in which he introduces the topic under the general heading of throwing weapons using the following words:

If you have to defend yourself against a sudden attack by an enemy, put some distance between you to ensure your safety. In this case use the small weapon called a *tobi dōgu* 飛び道具 (throwing weapon), known to the author as a *shūriken*. Samurai normally despised these and other throwing weapons, but they are most effective if used skilfully; the *shūriken* can do great damage if it hits someone's eye.[6]

The primary use of any small throwing weapon is therefore to create some form of diversion or delay. This is consistent with the use of them in an improvised fashion in historical texts, as Fujita freely acknowledges when he writes about other things that can be employed in this way such as Japanese women's traditional hair pins and the utility knives that samurai kept in their scabbards. However, Fujita also includes drawings of *bō-shūriken* and two simple varieties of what we now call ninja stars. One is cross-shaped and the other is an eight-pointed star. They are very different in design from what one sees in movies today and look like a number of straight darts welded together. Fujita shows in the same picture how *bō-shūriken* were best thrown using a spinning action, so it is no great conceptual leap to envisage two darts brought together in a cross shape and thereby spinning even better, and because of their absence from the old manuals it is very tempting to conclude that they are Fujita's own inventions. They would have required very little imagination compared to the ninja parachute. As to a possible motivation, perhaps Fujita's desire to make ninjutsu relevant to warfare in the modern age and his reference elsewhere to 'many things that can be used in the manner of *shūriken*' provide the clue.[7]

In fact, the evidence suggests that Fujita is actually illustrating for the first time in any modern publication something that really did exist. This may surprise the reader who knows that ninja stars are not found in the ninjutsu manuals, but that is because we are looking in the wrong place. As noted earlier, the shinobi's personal weaponry was of no relevance to the authors of *Mansenshūkai* and *Shōninki*. The true preserve of side-arms are the Tokugawa Period *jūjutsu* manuals, where various weird and wonderful weapons appear, some of which have made their way into the armoury of the ninja of today. Unlike the ninjutsu manuals, writings about personal combat tended to be associated with particular schools of martial arts which developed their own specialities, including prowess with particular weapons. One such was the Seigō-ryū, whose repertoire of *jūjutsu* uniquely included the forerunners of the ninja stars of today, and one of their (supposed) documents even provides an illustration of them. The star-like variety consists of two blades fastened together with a hinge like a pair of scissors.[8]

Yet this document (discussed below) cannot have been Fujita's source, because he makes no mention of any manuscript by the Seigō-ryū in his extensive bibliography for *Ninjutsu Hiroku*, and in any case Fujita's drawings are very different. Fujita's sources are therefore highly likely to have been actual objects, because in a book called *Shūriken* published not long after *Ninjutsu Hiroku* its author Naruse Kanki includes photographs of several genuine specimens that were owned by the last Tokugawa Shogun.[9] One of the Shogun's weapons is identical to the drawing in Fujita's book, so it is almost certain that real objects were Fujita's sources, not his imagination. Incidentally, Naruse also deals with *bō-shūriken* and adds the fascinating story that examples were taken to Paris along with other samurai weapons by a trade delegation sometime around the time of the Meiji Restoration. They were put on show in a museum and the heavy steel darts would eventually provide the inspiration for the first ever aerial weapon: the darts dropped from early aircraft during World War I before the introduction of exploding bombs.[10]

To add further to the mystery, some time subsequent to the publication of his *Ninjutsu Hiroku*, Fujita Seikō came across the document mentioned above in connection with the Seigō-ryū of *jūjutsu* and published it, giving it the title of *Iga-ryū ninjutsu kaki hi no maki* 伊賀流忍術隠火の巻 (The secret ninjutsu fire scroll of the Iga school), even though (as Nakashima Atsumi points out), it had no connection with either Iga or ninjutsu and is only partly concerned with fire.[11] The original manuscript (which appears to have been lost) is believed to date from the mid Tokugawa Period. Fujita's version has since been republished in facsimile by the Iga-ryū Ninja Museum with notes by Kawakami Jinichi, including a transcript of the original text into modern Japanese along with similar material also from the Seigō-Ryū. In an introduction Kawakami explains that the document was presented to the museum by Fujita's bereaved family in 1974.[12]

Four throwing weapons are shown on the last page of *Iga-ryū ninjutsu kaki hi no maki*, with none of the accompanying text that is otherwise found throughout the book. They are an *uchine*, a *bō-shūriken* and two ninja stars captioned as *jūji-shūriken* (cross-shaped) and *happō-shūriken* (eight-pointed) respectively. The drawings have a black dot in the middle suggesting that they may be hinged. Nakashima reproduces the two drawings in his book *Ninja o kagaku suru*.[13] He points out that many types of concealed weapons were employed in *jūjutsu* so the throwing weapons are in good company in the 'Fire Scroll', along with the *konpi*, a short chain with an iron ball at one end and a cross that looks vaguely like a *shūriken* at the other. A 30cm rope is attached at the cross end. The *konpi* is thrown around the blade of an assailant's sword. There is also the *kakute*, a small one-finger knuckle duster 'which would be useful in a scuffle' and two forms of concealed pistol. As they are fired by smouldering matches they are a far cry from a derringer, so their secrecy pertains only to their small size. Nakashima adds that it is a mistake to think of concealed weapons as being the preserve of the shinobi, even if popular works give that impression.[14]

These two unique drawings of ninja stars soon replaced the 'welded dart' model of *shūriken* in Fujita's imagination and acquired a life of their own during the 1950s thanks to Fujita's collaboration with Okuse Heishichirō. They reappeared first in Okuse's innovative booklet *Ninjutsu* of 1956, but there they are presented as photographs of real objects, which suggests that someone had them made based on the drawings, and what could have been more natural for the enthusiastic Okuse than to commission a local blacksmith to make some for experimental purposes? In *Ninjutsu* he describes their use as being to distract an opponent as well as being an assassin's weapon.[15] The original drawings were then copied for an illustration to the novel *Yagyū Bugeichō* by Gomi Yasusuke in 1957.[16] Shortly afterwards the ninja stars made an important conceptual breakthrough in the manga *Ninja bugeichō*, which was composed sometime after 1957, where they were drawn for the first time being thrown in anger.[17] Then came their first 'live appearance' on film. As with other photographs in Okuse's *Ninjutsu*, the ninja stars had become line drawings in Murayama's novel *Shinobi no mono* in 1962, the book for which Okuse provided expert assistance. That provided their entry into the movie of the same name for their debut on screen, and a similar sequence would follow in the television series *Onmitsu Kenshi*.[18]

Spurred on no doubt by the welcome publicity, Fujita returned to the subject of *shūriken* in a book of 1966 entitled *Zukai Shūriken-jutsu*. On the face of it this dense compilation of *shūriken* techniques is a very learned and fitting work for a master of ninjutsu because the book has page after page of carefully delineated drawings of the different types of *shūriken* associated with various schools of ninjutsu and *jūjutsu* and how to throw them. Yet thirty years have passed since *Ninjutsu Hiroku* and the needs of World War II are a distant memory. The ninja boom of the 1960s is now in full swing, with *shūriken* being flung in all directions, so *Zukai Shūriken-jutsu* unfortunately has to be seen as an attempt by Fujita to cash in on their popularity in the movies by writing a detailed manual about the weapon. The end result is very similar to Naruse's 1943 effort. Fujita begins with the *bō-shūriken*, and as for the fine detail of his spinning ninja stars, there is a vague suggestion that the *jūji-shūriken* may have been copied from the cross-shaped version of a *kongō* 金剛 (*vajra*, a symbolic thunderbolt used as a Buddhist ritual object to signify power). There is also the completely unhistorical (and highly uncomfortable) inclusion of a 'carrying-rack' for *shūriken* hung from a samurai's sword-belt,[19] although the most serious questions regarding invention relate to Fujita's picture of a *shūriken* shaped like a swastika.[20] The design appears in every modern book about ninja, even though its true origins were revealed in 2008 following an interview with the actor Ōse Kōichi, who played the samurai hero in the 1962 television series *Onmitsu Kenshi*. He states that the swastika *shūriken* was invented because of concerns that the pointed *shūriken* were too dangerous to use on set.[21] Yet in 1966 this television prop is being illustrated in an apparently learned book by Japan's supposed master of ninjutsu!

The design of the ninja stars would soon be elaborated further to make the vicious-looking fantasy weapons of today. The demand for them grew quickly, helped no doubt by the release of *You Only Live Twice* and Hatsumi's demonstration of them in the vignettes at the end of *Shōnen Ninja Kaze no Fujimaru*. In December 1967 an advertisement appeared in *Black Belt* offering ninja stars for sale from an address in Albuquerque, New Mexico.[22] These monstrosities and their descendants are highly dangerous weapons that have resulted in several eye injuries to children, yet ninja stars as a general category remain inseparable from even the highly sanitised fantasy ninja figures today who throw them at monsters and villains.[23]

The transformation of the ninja star from an obscure novelty weapon to its present status as the ninja's defining object sums up the notion of the ninja as an invented tradition, because Fujita Seikō did not invent the *shūriken*. Just as with so many other facets of the ninja myth the object itself was never a complete fabrication; it was the tradition that was invented when Fujita took a minor weapon from *jūjutsu* and placed it into the ninja's armoury along with the black costume and the water spider. So the answer as to why *shūriken* do not appear in the manuals is not that they did not exist, nor that they were secret weapons. The object was not recorded in the ninjutsu manuals because it was irrelevant to the subject matter and appeared in only one *jūjutsu* manual because it was an unimportant curio that the Seigō-ryū alone had explored, a footnote to the gentlemanly martial arts of the Tokugawa Period.

Now all has changed. In Iga City one can eat *shūriken*-shaped biscuits and throw blunt metal *shūriken* at targets outside the Tourist Information Centre. In 2017 the sponsors who signed up as associates to the Japan Ninja Council were given a folded paper origami *shūriken* as a memento, and because of the links currently being made to the 2020 Tokyo Olympics I have little doubt that we will see ninja stars being thrown during the opening ceremony. The stone the builders rejected has become the corner stone.

Chapter 14

Selling the Shinobi

In the introduction to this book the essence of the ninja myth was defined as a search for continuity, and throughout the long history of the myth's development one genuine example of continuity that crops up time and again is the use of the contemporary ninja figure to make money. Over three centuries ago the plays of the kabuki theatre sold the image of a ninjutsu magician to the eager population of Edo just as the theme parks of today sell a fantasy ninja whose identity one can adopt for a few hours of escapism. Regardless of what name he may have been given at any particular time in history, the ninja has always been a marketable commodity, although there have been situations where a possible ninja connection could have been exploited yet someone has deliberately chosen not to do so.[1]

Iga City lies at the heart of the modern ninja tourist industry thanks to the determination of Okuse Heishichirō. The lucky conurbation has never looked back from its modest beginnings in the 1950s and has retained a pre-eminence in all things ninja to this day. It still grasps any opportunity to enhance the brand, and in February 2017 Iga officially designated itself Ninja-shi (Ninja City), a public relations exercise that was inspired, according to its press release, by the fact that Kagawa Prefecture had declared itself 'Udon-ken' (Noodle Prefecture).[2] In April of the same year the 'Ninja Villages of Iga and Kōka' were given special heritage status by the Japanese government. Hampered only by poor train connections a stay at 'Ninja City' offers visitors a rich and very enjoyable ninja experience under the guidance of its boy and girl ninja mascots; the cute Ninjack (Ninjaku) and Shinobuchan.

The starting point for most tourists is a large purpose-built complex called the Danjiri Kaikan, which houses the floats used in the annual

Tenjin Festival in October. They are displayed inside a huge exhibition area along with costumed dummies of participants and video recordings of the event. The Tenjin Festival has nothing to do with ninja and there are no specific displays about the topic, but built into the same complex is Iga City's Tourist Information Office and Shop where almost anything involving ninja may be purchased and almost everything involves ninja. Here are sold bottles of ninja beer, plastic swords, the much-prized ninja toilet paper, blunt *shūriken*, ninja T-shirts for dogs, ninja pyjamas, ninja hoods for cats and ninja underpants. Outside the main door is a throwing range for *shūriken* and set-piece photography scenes, while nearby one can eat 'ninja noodles' with seaweed in the shape of a *shūriken*. It is here that you can hire ninja costumes to wear throughout your visit, and the sight of groups of people dressed from head to foot in black while walking around town raises only the occasional eyebrow.

Thus attired, the aspiring ninja begins the short walk up towards Iga-Ueno Castle, one of the first examples of the modern trend for rebuilding tower keeps in concrete. Iga-Ueno dates from the 1930s. Inside the castle are fine displays that concentrate almost totally on the Tōdō *daimyō*, while a few token dummy ninja hang from the walls in a melancholy fashion, but on walking down the rear slope from the castle your eye is caught by a fortuitously situated Inari Shrine. Its vivid tunnel of vermilion shrine gates has appeared in many a ninja documentary, and beside it is the attractive old farmhouse that is now Iga's ninja mansion. Some of the 'ninja' elements probably existed in the original building, but would not have had the same significance. A removable floor panel, once used perhaps for concealing valuables, now holds a sword. A stowaway ladder, a sensible inclusion in any restricted space, is presented to visitors as a secret staircase. Other more questionable features are included that have been copied in other ninja houses, such as the wall panel that opens by slipping a piece of paper into the door jamb. These elements are demonstrated by guides dressed as ninja including young lady ninja attired in startling pink. *Shūriken* are freely thrown against interior walls that have been thoughtfully encased in polystyrene foam.

On leaving the mansion the next stop is the famous Iga-ryū Ninja Museum, which is built into the side of the castle hill. Fujita Seikō's collection of ninjutsu documents were once on display, but are now in the care of Mie University. The Iga-ryū Ninja Museum provides a permanent showcase for Okuse Heishichirō's view of what a ninja was and did, and nowhere else in the world can you get a better idea of the original concept of the ninja as he is now known and loved. There is a wonderful quaintness about the place even though it was completely refurbished quite recently, and judging by my comparative observations made in 1986, 2004 and 2015 the opportunity has been taken to play down the more outrageous claims once made for so-called ninja equipment, although a firework-projecting tube is still labelled a 'ninja wooden cannon'. There is, however, an official denial that ninja ever wore black, for which the English version of the display card reads:

> Although today's common notion is that the ninja wore a black outfit, a black one is rather conspicuous because its outline becomes apparent on a night which is not totally dark. In reality, the ninja wore working clothes dyed navy blue worn by farmers...

Needless to say, the costumes worn by the guides, the display mannequins and many of the visitors are far from being farmers' working clothes, but this is just one example of the blend of entertainment with irritation that the museum presents. It succeeds to some extent in its aim of presenting the reality behind the ninja concept, although the fantasy ninja and the historical shinobi tend to get horribly mixed up and if there is a conflict the commercial needs triumph. The visitor may be even more baffled by attending the ninjutsu demonstration in the adjacent purpose-built area. It is very well done and highly entertaining, but it leaves one in such confusion about what was ever real about the ninja that a healthy dose of ninja beer becomes a vital necessity.

Few tourists venture beyond Iga City to see the other ninja-related places that Okuse almost literally put on the map. They are helpfully set out in a very useful English language guide-map prepared by local ninja enthusiast Ikeda Hiroshi and include the battlefields of the Iga Rebellion, old castle sites, important shrines and privately owned places like the Sawamura house.[3] You can also make a pilgrimage to the graves of the authors of *Mansenshūkai* and *Iran ki*. The Aekuni Shrine, once burned to the ground by Oda Nobunaga and the site of the attempted assassination of him, was rebuilt in 1593 and is well preserved. Just outside its main gate is a small colourful shrine to Kōga Saburō, the forefather of the famous Mochizuki family. The Aekuni Shrine was greatly honoured by the Tōdō family when they gained possession of Iga-Ueno Castle in 1608 because it was located to the north-east of the

175

castle. That was the 'devils' gate' from where evil could strike, so Aekuni acquired a considerable protective significance. Its long history and stunning vermilion colour make it well worth a visit. To the south of Iga City lies the Hanagaki Shrine. It was founded in 1004 and was associated both with the Tōdō's chief retainer Tōdō Uneme and the Hattori family, because the place enshrines the *ujigami* (the tutelary deities or 'household gods') of the Hattori. In the Mibuno area is the site of the castle of the same name that was attacked during the Iga Rebellion and the perfectly preserved house of Sawamura Jinzaburō Yasusuke, the shinobi who spied on Commodore Perry. It is not open to the public, but its outside appearance and subtle defensive layout have not changed in centuries.

One great advantage that Iga possesses over its rival Kōka is that Iga's main ninja features are concentrated within a small city area and may easily be accessed on foot. Kōka's ninja sites are more diffuse and can only be covered in their entirety by car using a pamphlet featuring a cute ninja mascot called Ninjaemon.[4] A bronze statue of a ninja stands outside Kōka Station, from where it is a short drive to Kōka's own ninja house. This one dates from the late seventeenth century and was owned by the Mochizuki family. For many years it was used for the production of pharmaceuticals, for which Kōka became renowned. Unlike the Iga ninja house it is also used as a museum with some interesting display cases in addition to the usual revolving doors and secret passages.

Kōka also boasts its own ninja village, a small-scale effort that is rather like a cross between an open-air museum and a historical theme park. It lies in an attractive rural area and is so well done that the two elements of history and enjoyment never seem to clash. It contains two very interesting old houses that once belonged to the Fujibayashi and Okada families and are set in an attractive wooded rural area. There are also fun things to do including walking across a pond while wearing water spiders and holding on to a rope. Combined with a tour of Kōka's shrines, a pleasant and informative ninja day may be had. The foremost shrine in the area is the Aburahi Shrine for Aburahi Daimyōjin, the tutelary deity of the families of Kōka and the *kami* of oil, so offerings of cooking oil and lubricants are still made at his simple yet beautiful shrine at the upper end of the Somagawa Valley. It houses a very interesting local history museum, which includes material on shinobi. Not far away is the Yagawa Shrine. There are historical records of the Kōka-gun Chūsō holding meetings at the Yagawa Shrine and a stone monument within the precincts commemorates this.[5] At the Buddhist

Jigenji is the *ihai* (memorial tablet) and a stone grave marker for the Kōka men killed during the defence of Fushimi Castle in 1600. Another place worth visiting is the Kōka Museum of Pharmacology. There are extensive exhibits dealing with the local area's tradition of medicine, including sections about Yakusojin the god of healing, but ninjutsu and ninja history are brought in too, although in quite a low-key manner.[6]

Apart from places like these ninja can pop up anywhere in Iga and Kōka. Cute cut-out wooden ninja are often placed at pedestrian crossings outside schools, but the great surprise for a historian is to find locations that could exploit a ninja connection, but choose not to do so. Two examples are Ishibe and Minakuchi, which contain a number of historical sites relating to important events described earlier in this book. In Ishibe is the Chōjūji, the beautiful mountainside temple from which Oda Nobunaga removed buildings to create the Sōkenji at Azuchi Castle, and the foundations of the stolen pagoda are clearly visible. The site of the Rokkaku's fortress of Ishibe Castle is now another temple, but no reference is made at either place to any link with Kōka's shinobi past. In Minakuchi City the hog's-back *yamashiro* of Minakuchi Okayama Castle is a prominent feature, yet again no connection is made between the old Kōka-gun and ninja. The same applies to almost the whole of the administrative area called Nabari City that shares Mie Prefecture with Iga, although Akame is an exception. There are long hiking trails through the forests from where the very fine series of forty-eight waterfalls may be viewed. The owners of the falls themselves make no connection with ninja, but a well-organised establishment within the village does, claiming that the Akame Falls were the site of centuries-old ninjutsu training. The claim is of course totally without foundation and is probably based on a tenuous connection to *yamabushi*, because they are known to have practised austerities such as standing naked under waterfalls. Needless to say, a great conceptual leap has been made to transform *yamabushi* into ninja, but that is enough to sustain a ninja training centre where visitors can dress up to tackle assault courses and throw *shūriken*.

The commercial exploitation of the slightest ninja connection is such a potential money-spinner that it is not surprising to find that places with no relation to either Iga or Kōka have got in on the act. To some extent this is a perfectly legitimate activity because of the legal requirement for every *daimyō* during the Tokugawa Period to maintain his own surveillance programme by men who were often called shinobi, so every prefecture in Japan has its own 'ninja connection' if it chooses to look for one. In the city of Kawagoe in Saitama Prefecture the privately owned Kawagoe Historical Museum has a small display of ninja equipment with a card describing their local shinobi tradition by way of justification.[7] A more recent example is in Hirosaki City in Aomori Prefecture. An old farmhouse has been discovered, which once belonged to a man employed by the territory's ruler during the Tokugawa Period as a *hayamichi no mono*, the local name for a shinobi. He even had links to Kōka, and the existence of trap doors and hiding places within his home means that its transformation to a ninja house cannot be far away. One can only imagine the delight that must now be felt within the Hirosaki Tourist Office.[8]

A long-established ninja connection (even though it is officially denied) may be found in the city of Kanazawa (Ishikawa Prefecture) where a temple called the Myōryūji possesses a number of secret rooms, concealed doors and secret passages, and is now inevitably known as Ninjadera (the ninja temple). One of its most interesting features is the outer entrance door, which is approached up a short flight of stairs. The vertical surfaces of the stairs are only paper, as used on sliding screens, so that a guard could thrust a spear through to cut the legs of anyone trying to gain access. The sliding entrance door is in fact a double one, the inner of which slides across a hinged door which leads to a secret corridor. A person being pursued could therefore give the impression of disappearing. There is also a completely hidden staircase to the third storey where there is a tea ceremony room. In the *tokonoma* (alcove) is a rather ugly painting of Mount Fuji, which is hinged, and drops down to let a guard through. The temple's well, which can be seen in the tiny courtyard, is supposed to have a secret passage leading to the castle, which seems rather unlikely as the river is in the way. All the guide books say that there is no connection with ninja, but the mere mention of the name even in denial has placed Myōryūji firmly on the Kanazawa tourist scene.

In Japan's former capital of Kyoto a number of places cover the entire spectrum of ninja-related connections from the studiously ignored to the blatantly commercial, although it is quite remarkable to see how Kyoto's great historical sites treat any potential ninja connection with calculated dismissal. The Ni-no-maru Palace of Nijō Castle has its famous 'nightingale floor', where carefully counterbalanced floorboards mounted on sliding metal hinges give warning of anyone's approach. Protection against assassination is mentioned in the guidebook, but there is no mention of ninja trying it out, and there is also no direct mention of ninja in connection with one of Kyoto's most interesting old buildings, a small urban house called Nijō Jinya. This seventeenth-century mansion has many of the security features found in the ninja houses of Iga and Kōka, but commercial exploitation in that regard is completely avoided. The place was thrust into tourist prominence in 1964 by a book called *Kyōto, a contemplative guide*.[9] The word *jinya* means 'encampment house', in other words a mansion that possesses defensive features. It was the home of Ogawa Hiraemon, an impoverished samurai who became a rice merchant during the early Tokugawa Period. His business flourished, and connections with the imperial court brought Ogawa into contact with the upper reaches of Tokugawa society. Visiting *daimyō* would choose to stay at his house, so over a period of years Ogawa became an innkeeper by default. As the landlord to so many distinguished guests Ogawa was naturally concerned for their safety and built into his establishment a number of defensive features to reassure them. The flooring of the entrance hall has a built-in squeak like the nightingale floor of Nijō Castle. Elsewhere in the house are removable floorboards disclosing ankle-breaking beams, and narrow hallways with low ceilings to discourage swordplay. In the main reception room there is a skylight window that provides access to a secret room from where a guard could drop down on to a suspicious visitor. The guard chamber also allows whoever is in there to hear all conversations in the room below. Ogawa was far ahead of his time in terms of fire precautions, which included doors inlaid with clay and metal-sheathed roof sections. These were so effective that they allowed Nijō Jinya to escape the great fire that ravaged Kyoto in 1788.

Even more downplayed than Nijō Jinya is the lack of ninja connection with a small temple to the south of Kyoto called Taikō-An. It lies near Tōfukuji and contains a tea room where the leaders of the Western Army planned the battle of Sekigahara in 1600. Just below the ceiling of the tea room is a skylight opening similar to that in Nijō Jinya, and the guide will tell you without batting an eyelid that Tokugawa Ieyasu concealed a shinobi there to listen to what was being discussed. Whether or not the story is true, a remarkable opportunity to market Taikō-An as a 'ninja temple' has been respectfully declined.

At the opposite extreme lies Kyoto's Ninja Labyrinth. This is a genuinely non-authentic ninja-themed restaurant with an attached souvenir shop where the waiters dress in black and take you to your table in one of a series of underground booths built like caves. Small groups experience the magic of ninjutsu by means of card tricks, while

179

larger groups dine in an annexe and enjoy a ninja floorshow. It is the Japanese equivalent of a mock medieval banquet and all you pay for is the food, unlike Kyoto's most expensive ninja venue, which is located within the Toei Eigamura (Toei Studios Movie Village). This provides an exciting day out for anyone interested in Japanese films and contains several excellent historical sets where samurai dramas are still filmed. As the whole place is of course artificial, a ninja village fits quite neatly inside it. The ninja section is extensive although an extra charge is made. There is a very good Ninja Trick House (essentially an indoor maze with concealed doors and passages) and Japan's largest ninja souvenir shop.[10] As for making guests feel welcome, not far from the Fushimi Inari Shrine is an old house now called the Ninja-dō, where one may dress up and act as a ninja. It thoughtfully includes a prayer room for Muslims.[11]

Tokyo has its own a ninja-themed restaurant, but because authentic shinobi once guarded Edo Castle, ninja tourism in Tokyo also has something of a serious side. Hattori Hanzō's name was given to the Hanzōmon, one of the gates of Edo Castle that his men guarded, and he is also now remembered in the name of a subway line. Hanzō's grave lies in the grounds of the Saienji temple, and his spear is preserved in the temple hall. The musketeer corps from Kōka are remembered at their barracks in the East Palace grounds of Edo Castle, and there are also a number of important Shintō shrines connected to the Iga- and Kōka-mono. The Onden Shrine in Harajuku enshrines the spirits of the Iga-mono who helped Ieyasu escape in 1582. The building is a modern reconstruction. The Kōka-mono have their own Inari Shrine in Sendagaya, which is also active and well-preserved, and under the arches of the railway bridge of Okuno Station in Shinjuku someone has painted a fine mural of the musketeer corps in action.

All these places should be on the itinerary of the ninja pilgrim who seeks a historical authenticity that goes behind the myth, but it is also hard to resist the tendency simply to enjoy places like Toei Eigamura that throw up into the air any pretence towards a genuine ninja connection and wallow in the profits from the copyright-free image of the man in black. To add to the ninja restaurants and film studios most of Japan's samurai theme parks have ninja sections with floor shows and mock ninja houses based on the Iga original, and several new ninja theme attractions have sprung up in the last few years. Ureshino, a rather run-down spa town in Saga Prefecture, now has its own ninja village that may compensate in some small part for the closure in 2014 of one of Japan's last sex museums.[12]

A long-established ninja presence is to be found in Togakushi, a village in Nagano Prefecture where two ninja attractions provide income for the local economy when the ski season is not in operation. The older of the two complexes owes its existence to Hatsumi Masaaki, a native of the area, whose own Togakure-ryū (the pronunciation of the name Togakushi in its ninja context) school of ninjutsu claims a pedigree dating back to a retainer of the twelfth-century samurai Minamoto Kiso Yoshinaka. The man, called Nishina (or Togakushi) Daisen, is supposed to have fled to Iga, where he learned ninjutsu. He then returned to Togakushi to pass on his new knowledge. The museum within the complex is almost a shrine to Hatsumi, whose vast personal collection of ninjutsu equipment and tools is displayed along with numerous photographs of him in action. There is also a *shūriken* throwing range and a very complex and puzzling ninja fun house.[13] A short drive away lies the Chibikko Ninja Village, where the emphasis is much more upon the younger visitor. There are extensive play areas and two fun houses, one of which is child-sized, but there is also a museum that is every bit as interesting as the Hatsumi version. The Chibikko displays lay an emphasis on ninja ephemera with a large and very well organised display of ninja-related movie posters, manga, books, toys and souvenirs. The objects date back many decades, but are also up to date, as shown by the extensive collection of *Ninjago* figures.[14]

The two Togakushi places have certainly set a standard for Japan's newest ninja village: the Shinobi no Sato, which opened in 2015 at Oshino 忍野 in the foothills of Mount Fuji, a location whose name fortuitously shares the character *nin*. No historical ninja connection is claimed for its location, which can only have been chosen for its proximity to other tourist traps in the Mount Fuji area, nor is any attempt made to educate the tourists who are bussed in to throw *shūriken* and visit a rather dull ninja fun house.[15] There is no museum and very little in the way of explanation as to why a small-scale theme park should be staffed by people dressing in black. At Oshino the ninja have lost any semblance of historical reality. Divorced from Iga and even from history itself, they exist only in the ruthless world of commercialism.

Chapter 15

The Exemplary Ninja

Since 2012 there have been a number of wide-ranging developments in the ninja myth's long history of reinvention. At one extreme they involve the serious academic study of the ninja's ancestry and the application of scientific methodology to what the manuals of the Tokugawa Period understood as ninjutsu. At the other end of the spectrum we find a high moral dimension based on an idealised view of the ninja as a behavioural and physical exemplar for Japan's youth. Both these developments embrace lofty goals and have noble Confucian overtones yet, as in so many other previous examples of ninja-related activities, they are underpinned by hard-nosed commercialism.

The commercial aspect is best illustrated by how ninja-related innovations are publicised by the Ninja Council of Japan, the official body established in 2015 to bring a range of disparate activities under one coordinated umbrella. It has a website and a daily Twitter feed, where the widest possible range of ninja-related activities, from the deadly serious to the utterly ridiculous, are presented in an unbiased fashion.[1] Among the new ventures the ones that have attracted most overseas publicity are directly related to the build-up to the Tokyo Olympics of 2020. When it comes to exciting popular imagery to publicise a major world event samurai are clearly not enough, and a mere glance at the pages of a ninja manga must have convinced a sober-suited official that the warriors in black could be a great lure for tourists. The first activity carried out under this heading was the appearance in 2016 of advertisements recruiting people to act as ninja to entertain visitors to Japan. Nagoya Castle in Aichi Prefecture was one of the first off the mark, advertising for six 'ninjas' (sic) to be trained for work during the Olympic period. Aichi's association with ninja has some historical validity because of the service Kōka rendered to the future Shogun Tokugawa Ieyasu in Mikawa Province, but the commercial exploitation of a ninja connection appears to be a recent development.[2]

To some extent the linking of ninja to the Olympics is a natural progression from a related trend that had been underway long before the advertisements were placed. This was the use of ninja as physical and moral exemplars for the nation's youth. They were seen to be ideal for this purpose for two reasons. The first was their legendary athletic prowess, which could be used to promote physical activity and an interest in sport. The second was their exemplary code of loyalty and their steadfastness, derived from the *nin* of endurance. To some extent this is no more than the modern application of a centuries-old tradition of finding spiritual values within the martial arts, and to apply this idea to ninja has echoes of Itō's and Fujita's work. Yet those two authors lived during a time of war, and the obvious weakness in using the ninja in this way in a time of peace is the fact that the slightest investigation into the ninja's historical forebears reveals that they were trained in these high-minded skills to enable them to break into buildings and cause havoc. Once again Jonathan Clements sums up the situation perfectly. He is writing about the overall modern interpretation of ninjutsu, but his comments can be applied directly to the new initiative:

> Ninjutsu's other big problem comes from its purported origins in a toolkit for assassination. Judo and the other martial arts have struggled for a century to establish themselves as philosophical exercises, as mental regimes that soothe and tame savage thoughts. Ninjutsu pays lip service to this ideal… but at the most basic level ninjutsu's claim for authenticity has forced it to acknowledge that certain of the manuals are concerned with breaking and entering, poisons and sneak attacks.[3]

Quite clearly, in order to become an example to the young and a 'soother and tamer of savage thoughts' the ninja needed to be sanitised for our protection, so the ninja's ruthless antecedents were quietly put to one side in place of a figure who could be at one and the same time both noble and unbearably cute. The comic book fantasy ninja therefore became a role model in place of the dedicated infiltrator, and the results were unveiled in some style by the launching of an exhibition for children called 'The Ninja: Who Were They?' (*Ninjatte nanja*) in 2016. The event was first held at Miraikan, Tokyo's 'Museum of Evolving

Science', and has since become a travelling roadshow, visiting Miyazaki Prefecture in the summer of 2017. It takes itself very seriously, promising to treat ninjutsu as something learned, scientific and health-giving, as its website indicates:

> The Ninja were people with comprehensive abilities in terms of 'mind, skills and body' and ninjutsu was an accumulation of practical knowledge concerning nature and society. This exhibition offers you the opportunity to train yourself in throwing shuriken, improving your jumping power, and introduces skills like memory enhancing techniques, sending secret messages, and special breathing techniques necessary for successful completion of a mission. By studying old ninjutsu manuscripts, and with the help of modern science, we can get close to the 'True Ninja' and find hints that may help us to survive in the coming future.

The references to jumping and breathing have obvious echoes of Fujita Seikō's claims to extreme physical prowess, but the exhibition's young visitors were spared any requirement to thrust hat pins through their cheeks or eat wine glasses. Such austerities were replaced by scientific data about the improved respiration that came about when one trained in 'the ninja manner'. The exhibition also reflected in a very forceful way the view that ninja should be officially regarded as having really existed. Indeed, the impression was given most strongly that the ninja still existed in the form of modern practitioners of ninjutsu, so Kawakami's definition of a ninja as a fictionalised shinobi found no place in the Miraikan. Instead the historical continuity envisaged in the ninja myth was accepted so completely that no practical distinction was made between past and present or even between fact and fiction in the pursuit of its lofty goals.

The official handbook for the exhibition is a revealing piece of work that sets out these ideas to the accompaniment of attractive illustrations of the acceptable ninja in action. If the word shinobi appears in the text it is used only as an adverb to describe what was done by the historical ninja, whose authenticity is never questioned. For example, a map and accompanying table showing the words used during the Sengoku Period for those who carried out undercover work such as *rappa* and *suppa* is presented as a guide to alternative names for ninja. There are also some classic statements in the book's English summary, such as 'Ninja sought to live in harmony with nature, worked without fanfare on behalf of their communities, and were resilient in overcoming hardship to survive. [Ninja were] possessed of the mental fortitude to stay calm in any situation'. One longed for the slogan, 'Ninja Keep Calm and Carry On', a missed marketing opportunity if ever there was one.[4]

The young visitor to the exhibition experienced a succession of three zones: 'Perfect your mind', 'Enhance your skills' and 'Improve your Body'. A certificate ('valid as long as you live as ninja and non-transferable') was then presented and the child departed through the gift shop. It contained an exhortation to good health and service to the community of which Itō Gingetsu would have approved:

> Requirements for the ninja: Improve your body, skill and mind and maintain a good balance.
> Body: Build your physical strength and keep in good health.
> Skill: Acquire the ability to understand the environment without relying on the internet.
> Find a mission for yourself and live through anything to the bitter end.
> Protect the community you belong to.

The child was then given a severe personal and societal obligation:

> Since you have learned what the true ninja are and have been certificated as a ninja to live in modern society, you are now expected to complete special missions. It is up to you which missions you will engage in. Think that every day is another opportunity for you to improve yourself and live with the wisdom of the ninja.

The Miraikan event therefore illustrated once again the inherent tension in presenting the ninja as either fact or fiction. Just as in the case with my own training manual, the moral and health-giving exhortations at the Miraikan depended upon a suspension of disbelief in the reality behind the ninja myth.

The ninja goes to college
Meanwhile, Mie University's Iga Ninja Culture Collaborative Field Project continues to plough a solitary academic furrow through the world of the ninja. One of its aims is to study the ninja in a scientific rather than a historical manner, applying the techniques of sports science and human physiology to their allegedly superhuman

achievements and, at a more mundane level, to an examination of the reality behind the devices included in the Tokugawa ninjutsu manuals. Nakashima Atsumi (who is not directly connected to the Mie project) has been a pioneer in this field for several years. His *Ninja o kagaku suru* is an earlier attempt to apply the scientific method to ninjutsu,[5] although his greatest strength lies in the overall analysis of the Tokugawa ninjutsu manuals. He has produced the critical edition of *Mansenshūkai* and a shorter book in which all the main topics from the 'big three' manuals are studied side by side.[6]

The initial studies carried out by Mie University go beyond Nakashima's work by applying experimentation to topics in the ninjutsu manuals ranging from botany to explosives. The first results were published in a book in 2017. These include a detailed study of the use of smoke signals based on wolf dung involving combustibility tests, distance measurement and so on. There is also a memorable study of the water spider that brings in physics and hydrostatics. Some complex mathematical equations to do with upthrust are presented, which may even prove that the two-legged version of the water spider could actually work![7] Another important contributor to the field is Komori Teruhisa, whose own research interests include breathing techniques and the psychology of relaxation. He has studied the effects on relaxation of the breathing techniques recommended in the ninjutsu manuals, and through the concentration involved in performing the nine hand signs.[8]

Mie University also has a considerable programme of outreach through public lectures, and this already has an international dimension. Its most active overseas ambassadors are Professor Yamada Yūji and his colleague Kawakami Jinichi. They form the perfect combination, because between them they personify the traditional Japanese values of respect for authority and respect for ancestry. A typical overseas presentation involves a lecture by Yamada on the history behind the ninja phenomenon, followed by a demonstration of simple ninjutsu techniques by Kawakami based on material in the manuals rather than martial arts. Their written output is also considerable and gives a good idea of where these two key individuals think the ninja phenomenon might be heading in the twenty-first century, but once again there is an undercurrent of tension between the interpretations of the meaning of *nin* as either secrecy or endurance.

Kawakami Jinichi presents his personal understanding of the ninja phenomenon in his 2016 book *Ninja no Okite*. He begins by describing his own background and how he began his martial arts training at the age of six.[9] He is descended from the Ban family of Iga, so it is almost inevitable that any media coverage of him labels him 'The Last Ninja', although his first encounter with the word ninja came through watching a movie at the age of thirteen.[10] He evidently embraced the concept very rapidly, and as Honorary Director of the Iga-ryū Ninja Museum he exercises great influence over the brand. In 2012 he was appointed Honorary Lecturer at Mie University as part of the ninja research project and has made a major contribution to its work. The appointment attracted some interesting media attention at the time including a curious reaction from China, which was published in a revealing article entitled 'Does Japan want a revival of ninja culture?' The feature began with a brief summary of ninja and their Chinese antecedents. After noting that to be a true ninja one required hard training and discipline, the article developed into an attack on what it saw as decadent Japanese culture:

> Ninja culture nowadays can only be seen in cartoons and movies and the spirit of ninja is rapidly fading. The environment that surrounds younger people today values individualism, fads and identity. There are junior high school students who date older people in order to receive money in return, there are passive and weak 'herbivorous' boys as well as 'feminine' boys who have effeminate mannerisms and behaviour. The spirit of ninja is nowhere to be found in modern Japanese society.[11]

The author clearly regarded Kawakami's appointment as an indication of a revival in Japan of a hard masculine ninja culture and went on to praise the ninja's legendary fidelity, loyalty and calmness.[12] What is missing in the Japanese context, of course, is any notion of the ninja being a counter to a plague of gay vegetarians, so the article tells us less about ninja and more about a particular Chinese view of Japan. As for Kawakami's own ideas, he begins *Ninja no okite* with the problem of the self-evident lack of a role for the ninja in modern society:

> I who was born in this present age have been called the last ninja. However, to tell the truth, I am not a ninja. Some people come up to me and say, 'I want to be a ninja', but nowadays learning the techniques of entering houses in secret, crawling through tiny holes and concealing one's person is not something you can live off. Under this definition there is no need for ninja in the modern world.[13]

187

Kawakami's answer to the conundrum of the skills of an ancient sneak thief being in any way valued or useful in the modern world is that of seeing the ninja as an exemplary figure, so once again we encounter the spiritual claims that have permeated martial arts from the Tokugawa Period onwards. The mastery of ninjutsu requires daily training (*tanren* 鍛錬) without a break, and *nintai* is what makes it possible'.[14] Kawakami also stresses the basic underlying spirit (*seishin* 精神) of the ninja and how their training methods and culture may bring benefits to the world of today, just as Fujita once sought to apply them to Japan at war.[15] In a section entitled 'The basis of shinobi is the spirit of peace', he uses imagery of the character *nin* on a red circle to symbolise harmony.[16] The principle of the righteous heart found in *Mansenshūkai* lies at the centre of the basic spirit of the true ninja, and is explained by Kawakami by using familiar words conveying endurance and patience. Yet Kawakami goes much further than this, by adding *wa* 和 (peace) and *itsukushimi* 慈しみ (love) to the ninja concepts.[17] 'As for what is called ninjutsu', he writes, 'these techniques have been brought together not to bring about wars but to achieve harmony'. The Japanese people, he adds, have this as their inheritance.[18] To Kawakami historical ninjutsu was a practical method for avoiding battle and bringing about peaceful and stable coexistence, so ninja were by no means dark warriors of dubious integrity, as is often imagined.[19] As for the ninjutsu of today, the wish of Kawakami Jinichi, the so-called 'last of the ninja', is that peace should be spread throughout the people of the entire world.[20]

It is interesting to compare this attitude to the views of Itō and Fujita, who expanded them to cover the needs of a nation at war. Kawakami is not too different. In his embrace of the notion of a ninja as an exemplary figure, Kawakami's own *nin* of endurance has been expanded to cover the needs of a nation at peace, a notion that is not too far removed from the purple prose of Stephen Hayes, the Western world's doyen of ninjutsu during the 1980s, who wrote that the ninja 'is in truth the bearer of the universal light in a world that sometimes grows dim'.[21] Yet these concepts are so different from the ideas that lay behind authentic 'robbers from Iga' going into battle in 1582, that one is forced to the conclusion that the idea of the ninja as the bringer of peace is as fantastic as the idea of a ninja who can fly, because once again we are faced with the harsh reality of the shinobi's historical antecedents. From Sun Zi to Fujita Seikō the rationale behind the use of intelligence gatherers and infiltrators is always hard-headed military tactics, not brotherly love. The *Art of War* may urge that successful espionage ensures that a general can avoid unnecessary bloodshed when 'those skilled in war subdue the enemy's army without battle', but this is by no means an altruistic aim. Battle is avoided in order to destroy your enemy more efficiently; the goal is not the achievement of peace, it is the achievement of victory.[22]

In *Ninja no Rekishi* Kawakami's colleague Yamada Yūji adds his own views to the topic of the ninja's modern transformation into an exemplary figure a notion that even creeps into his ninja training manual for today's children, where boys and girls who behave like true ninja are found to be doing good deeds such as offering their seats on the bus to senior citizens and sweeping the school yard. He is more realistic than Kawakami and holds back from embracing the ninja as a symbol of peace. Yamada first notes that outside Japan there has always been a strong tendency to link ninjutsu to the martial arts, yet there has also been an apparent understanding that the notion of a ninja transcended the martial arts because it possessed a mystical element. This mystical context is the same as that embraced by the earlier writers to whom the '*nin*' in ninjutsu meant the '*nin*' in *nintai*, so once again we have returned to this vital primary definition of *nin* as endurance, perseverance and fortitude. Yamada defends this exemplary idea while avoiding any extension towards brotherly love. Instead he argues that these traits sit very well with the image that contemporary Japan wishes to project of itself. Within the wisdom and resourcefulness of the ninja lie important characteristics that have been associated with the development of modern Japan: skilfulness, industriousness, organisation and perseverance.[23] In that sense the ninja can indeed act as a peculiarly Japanese moral exemplar for the nation as a whole, and may well contribute to its economic well-being. After all, the 1964 Tokyo Olympics played an important role in the development of Japan's economic boom, and there is little doubt that hopes are being entertained that the 2020 Olympiad will also deliver the goods, a national aspiration in which the kick-starting of a ninja boom may also play a part. This, in summary, is the ninja's new task as 2020 draws near. It is an interesting situation that strengthens ever further my thesis that the ninja myth has constantly reinvented itself over a period of many centuries and is continuing to evolve even today.

The ninja clamjamphrie

This book has suggested several models for the ninja's origins as ancient Chinese spies, lower-class Japanese criminals, independent *jizamurai*

189

from Iga and Kōka, special forces in a *daimyō*'s regular army, palace guards who specialised in intelligence work, spiritual exemplars, masked assassins or fictional magicians. The reality of the situation is that all are correct because the ninja is an invented tradition and its inventors could and did choose whatever they wanted.

But what an invention it is! The ninja myth is a complex and dynamic entity that is still evolving today. Itō, Fujita, Okuse and Hatsumi may have exercised their imaginations during the twentieth century, yet they and all the other contributors to the ninja myth drew upon genuinely ancient elements that were already in place by the end of the seventeenth century. Those accounts were largely based on solid fact, and I believe that there is so much that is historically authentic among the ninja's antecedents that the invention of the ninja tradition lies less in the creation of imaginative elements than in an inspired blending of genuine ones. Hundreds of years before the Tokugawa Period *Taihei ki* had suggested that the word shinobi could be used as a noun in addition to its usual employment as an adverb. The Jesuits' *Vocabulario* of 1603 confirmed this beyond any doubt, while the recently discovered Amagoisan document shows that the noun form was already in use in Iga – the heartland of the ninja myth – twenty years earlier. The significance of the word shinobi may have changed during the Tokugawa Period when there was no longer any need to enter castles under the cover of darkness, but the use of men with that title for detective work and peacetime intelligence-gathering by and among the *daimyō* shows that an undercover role continued among people who hailed from Iga and claimed a speciality for themselves. A number of important strands were developing that needed only a little touch of magic to bring them all together.

It is for this reason that the ninja myth is much more than just another example of an invented tradition in terms of Hobsbawm and Ranger's definition, although it has many features in common. Hobsbawm's phrase 'an attempt to establish continuity with a suitable historic past' describes precisely what Okuse did in his insistence upon Iga as the ninja heartland. Hobsbawm also noted how in an invented tradition 'ancient records have been reinterpreted and exaggerated to reinforce a highly localized understanding of a phenomenon'.[24] The best example of this is the modern understanding of the attack on Kasagi Castle in 1541, which has been interpreted as evidence that hereditary specialists from Iga acted in a mercenary role. Hobsbawm also identified the 'use of ancient materials to construct invented traditions of a novel type for quite novel purposes'.[25] In this way the idea of ninjutsu changed radically from being something magical, to a martial art that could be learned through rigorous training. The important shift in the use of the word shinobi from an adverb to a noun also 'extends the old symbolic vocabulary beyond its established limits'.[26]

Yet the ninja is more than just the sum of his borrowed parts. The benchmark ninja who was woven from these different strands is a dynamic creature with a long and very complex ancestry and an extremely composite nature, so from my point of view the best way of summing him up is to borrow a splendid word from Scots and call the ninja a clamjamphrie, an expression which can mean anything from an eclectic collection of artwork to a riotous assembly. It is strangely appropriate for a visually striking warrior figure created by an amalgam of history and fiction over a period of many centuries.

Which brings us to the ninja of today, a popular character who is responsible for a huge turnover in tourism and entertainment and whose existence cannot be ignored in a quest for the authentic shinobi. To remove the fantasy figure from 'serious' ninja endeavours would be like trying to remove Santa Claus from Christmas, and as long as his origins are explained a touch of fantasy is surely acceptable. That, at any rate, is my own defence for writing a training manual for ninja that will contradict just about every point I have made in this book. The commercial aspect of 'brand ninja' alone would fully justify the academic attention it is currently receiving at Mie University, although I appreciate fully that in this book the fantasy ninja's history has only been half told. First, there has been insufficient space here to examine fully the numerous and wide-ranging interpretations of the ninja clamjamphrie that have arisen outside Japan since the early 1970s, who can range from gunmen assassins to mutant turtles. Secondly, I have given little attention to his emergence in pop culture within Japan itself. The transformation of the ninja image in more mainstream fields has been enough for me, but I appreciate that a youth-orientated gap exists. To give one example from this complex genre, *Naruto*, one of the most popular anime and manga series of all time, has produced sales numbered in millions.[27] The character in *Naruto* is a youth described as a ninja even though he has abandoned the black costume for orange, showing how much the concept has evolved since the days of *Shinobi no mono*. As so many studies say, 'more research is needed'. Mie University's International Ninja Research Centre has a lot of work to do, and any aspiring researchers can count on my full cooperation.

191

In conclusion, the ninja myth, which has its origins in authentic accounts of undercover warfare in Japan, has reinvented itself over the course of many centuries, and never has this been more true than in the time since World War II. When Japan emerged into peace and prosperity a young and fresh consumer market called for heroes of its own. An appropriate response was made, and in a burst of creative energy and even more creative marketing the ninja myth entered its crucial phase as out of the silver screen stepped a figure whose existence had been hinted at for centuries.

So definitive was this process that the benchmark ninja depicted in the 2017 movie *Shinobi no Kuni* differed little from his black-clad predecessors in the film *Shinobi no mono* from fifty-five years earlier. Once again a proud bunch of underdogs from Iga Province (where else?) defied the ruthless Oda Nobunaga using almost every weapon in Okuse's armoury. Their *shūriken* may have been computer-generated, but twenty-first century ninja combat is reminiscent of Kikuoka Jōgen's heroic descriptions in *Iran ki* in everything except for a small and fully acceptable addition of fantasy. I am sure that Kikuoka would have loved *Shinobi no Kuni*! He may have been a little puzzled at first to find his local heroes dressed entirely in black, but once he recognised them as the shinobi of Iga a great familiarity would have dawned upon him. The end product – now usually called a ninja rather than a shinobi – (and why should the name not be read using its *on* form?) apparently wore a mask as a matter of course and threw strange weapons about, but Kikuoka would have felt very much at home in a world of exaggeration, which in itself may provide the clue to understanding the ninja myth. The ninja, whatever name he may have been given at any moment in time, has always been larger than life, and through this crucial factor of exaggeration one can perhaps identify the direct link with the Tokugawa Period that the ninja myth seeks so desperately.

Floreat Ninja! Long live the supreme martial artist who can be either good or bad depending upon the screenwriter's whim! May he (and she) always be infinitely malleable, changing social class, gender, species, age and nationality as readily as Jiraiya changed himself into a toad: the eternal, international, flexible and profitable clamjamphrie that even the cats in my garden recognise as a ninja.

Notes

Preface

1. Turnbull, Stephen (1991). *Ninja: The True Story of Japan's Secret Warrior Cult*, (London, Firebird Books).
2. Turnbull, Stephen (2003). *Ninja: AD 1460-1650* (Oxford, Osprey Publishing Ltd); Turnbull, Stephen (2008). *Real Ninja* (New York, Enchanted Lion Books).
    Turnbull, Stephen (in preparation). *Ninja: The (Unofficial) Secret Manual* (London, Thames and Hudson).
3. http://www.mie-u.ac.jp/topics/kohoblog/2017/07/post-1410.html (Accessed 21 July 2017).
4. http://www.human.mie-u.ac.jp/kenkyu/ken-prj/iga/kouza.html (Accessed 9 June 2017).
5. Their bibliography so far is: Yoshimaru, Katsuya; Yamada, Yūji; Onishi, Yasumitsu *et. al.* (2014). *Ninja bungei kenkyū dokuhon* (Tokyo, Kasamashoin); Yamada, Yūji. (2014). *Ninja no Kyōkasho: Shin Mansenshūkai.* (Tokyo, Kasamashoin), available in English as *The Ninja Book: the new Mansenshūkai* (Tsu City, Mie University) E-book; Yamada, Yūji (2014). *The Spirit of Ninja: A Study of the Global Ninja* Craze (Tsu City, Mie University) E-book; Yamada, Yūji (2015). *Ninja on Kyōkasho 2: Shin Mansenshūkai.* (Tokyo, Kasamashoin); Yamada, Yūji (2016). *Ninja no Rekishi.* (Tokyo, Kadokawa Shoten); Kawakami, Jinichi (2016). *Ninja on okite* (Tokyo: Kadokawa Shoten); Various Authors (2016). *The Ninja - ninjatte nanja!? - kōshiki bukku* (Tokyo, Asahi Shimbun); Yoshimaru, Katsuya & Yamada, Yūji (2017). *Ninja no tanjō* (Tokyo, Kikkaku); Various Authors (2017b). *Ninja no Dokuhon* (Tokyo, Takarajimasha).
6. Yamada, Yūji (2015). *Ninja Shugyō Manuaru.* (Tokyo, Jitsugyō no Nihon Sha); Yamada, Yūji (Ed.) (2017). *Ninja ninjutsu chō hiden zukan.* (Tokyo, Nagaoka Shoten).
7. Yamaguchi, Masayuki (1963). *Ninja no Seikatsu* (Tokyo, Yūzankaku). It was reissued in 2015 as *Shinobi to Ninjutsu*.
8. He has published Lemagnen, Guillaume (2015). *Demystifying Ninjutsu, a Necessary Task* (Amazon).
9. Clements, Jonathan & McCarthy, Helen (2006). *The anime encyclopaedia: a guide to Japanese animation since 1917* (Berkeley, Stone Bridge Press); Clements, Jonathan

(2013). *Anime: A History* (London, Palgrave Macmillan); Clements, Jonathan (2016). *A Brief History of the Martial Arts: East Asian Fighting Styles, from Kung Fu to Ninjutsu* (London, Robinson).

10. http://vintageninja.net/ (Accessed 27 February 2016).

11. Turnbull, Stephen. (2014). 'The Ninja: An Invented Tradition?' *Journal of Global Initiatives: Policy, Pedagogy, Perspective: Interdisciplinary Reflections on Japan Vol. 9 No. 1, Article 3*, pp. 9-26.

12. Cummins, Antony and Minami, Yoshiie (2011). *True Path of the Ninja* (Rutland VT, Tuttle); Cummins, Antony and Minami, Yoshiie (2012). *The Secret Traditions of the Shinobi* (Berkeley, Blue Snake Books); Cummins, Antony and Minami, Yoshiie (2013a). *The Book of Ninja* (London, Watkins Press); Cummins, Antony and Minami, Yoshiie (2013b). *Iga and Koga Ninja Skills* (Stroud, The History Press).

Chapter 1

1. The important role of the ninja figure within 'Cool Japan' is stressed in a number of recent works such as Yamada, Yūji (2016) *Ninja no Rekishi* (Tokyo, Kadokawa Shoten), p. 8; Kawakami, Jinichi. (2016) *Ninja on okite* (Tokyo: Kadokawa Shoten), p. 200 and Takao, Yoshiki (2017) *Ninja no Matsue: Edo-jō ni tsutometa Iga-mono tachi* (Tokyo, Kadokawa Shoten), p. 6.

2. For the original story see Aston, W.G. (1972). *Nihongi: Chronicles of Japan from the Earliest Times to A.D. 697* (Volume I and II) (Reprint). (Rutland,Vermont, Tuttle), p. 201; for an interpretation of Prince Yamato as a user of ninjutsu see Fujita Seikō (1936) *Ninjutsu Hiroku* (Tokyo, Chiyoda Shoin), p.11.

3. McCullough, Helen Craig (1959). *The Taiheiki: A Chronicle of Medieval Japan.* (New York, Columbia University Press), p. 185.

4. For example, Okuse Heishichirō includes Fujiwara Chikata in his 1963 book *Ninjutsu, sono rekishi to ninja* (Tokyo, Jinbutsu), pp. 48-49.

5. This is *Buke Myōmokusho* of 1806. See Sasama, Yoshihiko (1968). *Buke Senjin Sakuhō Shūsei.* (Tokyo, Yūzankaku), p. 85.

6. McCullough, Helen Craig (1959). *The Taiheiki: A Chronicle of Medieval Japan.* (New York, Columbia University Press), pp. 47-49.

7. Yamada, Fūtarō (2014a). *Kōka Ninpōchō* (Tokyo, Kōdansha).

8. For a clip of this charming creation see https://www.youtube.com/watch?v=GbRZ9eE_0lg. (Accessed 12 March 2016).

9. Nelson, Andrew N. (1997). *The New Nelson Japanese-English Character Dictionary; based on the Classic Edition by Andrew N. Nelson, completely revised by John H. Haig.* (Rutland, Tuttle), p. 428.

10. Manser, Martin H. (2011). *Concise English Chinese Chinese-English Dictionary 4th Edition* (Beijing, Xianggang, Shang wu yin shu guan), p. 376.

11. Schuessler, Axel (2007). *ABC Etymological Dictionary of Old Chinese* (Honolulu, Hawaii University Press, 2007), p. 441. I am indebted to Jonathan Clements for explaining all this to me.

12. Shibata, Renzaburō (1955). *Sarutobi Sasuke* (Osaka, Tachikawa Bunmeidō), p. 12.

13. Itō Gingetsu (1937). *Gendaijin no Ninjutsu translated with notes and additional material by Eric Shahan* (Amazon, 2014).

14. Shirato, Sanpei (2009). *Ninja bugeichō:Kagemaru-den 1* (Tokyo, Kogaku-kan Creative), p. 236.

15. Yamada, Yūji (2015). *Ninja no Kyōkasho 2: Shin Mansenshūkai*. (Tokyo, Kasamashoin), p. 4.
16. From a private conversation with the author on 27 February 2014.
17. Various Authors (2016). *The Ninja - ninjatte nanja!? - kōshiki bukku* (Tokyo, Asahi Shimbun).
18. Sekka, Sanjin (1914). *Sarutobi Sasuke* (Osaka, Tachikawa Bunmeidō), p. 1.
19. Shibata, Renzaburō (1955). *Sarutobi Sasuke* (Osaka, Tachikawa Bunmeidō), p. 12.
20. Itō Gingetsu (1909). *Ninjutsu to yōjutsu translated with notes and additional material by Eric Shahan* (Amazon, 2014), p. 28.
21. Adachi, Ken'ichi (1957). *Ninjutsu* (Tokyo, Heibonsha): Okuse, Heishichirō (2011) *Ninjutsu, sono rekishi to ninja* (Tokyo, Jinbutsu).
22. Doi, Tadao (Ed.) (1960). *Nippo jisho: Vocabvlario da lingo de Iapam*. (Tokyo, Iwanami Shōten), p. 606. For the history of the *Vocabulario* see Cooper, Michael (1974) *Rodrigues the Interpreter: An Early Jesuit in Japan and China* (New York, Weatherhill), pp. 222-223.
23. Yoshimaru, Katsuya; Yamada, Yūji; Onishi, Yasumitsu *et. al.* (2014). *Ninja bungei kenkyū dokuhon* (Tokyo, Kasamashoin), p. 36.
24. Yamada, Yūji (2016). *Ninja no Rekishi*. (Tokyo, Kadokawa Shoten), pp. 24-25.
25. Yamada, Yūji (2016). *Ninja no Rekishi*. (Tokyo, Kadokawa Shoten, pp. 26-27.
26. Kuwata, Tadachika. (Ed.) (1965). *Shinchō-Kō ki*. (Tokyo: Jinbutsu Ōraisha), p. 249; Elisonas, J.S.A. & Lamers, J.P. (Trans. and Ed.) (2011) *The Chronicle of Lord Nobunaga by Ōta Gyūichi*. (Leiden: Brill), p. 314. The latter authors note that these fortresses were located in the north of what is now the modern city of Kobe in Hyōgo Prefecture.
27. Kuwata, Tadachika. (Ed.) (1965). *Shinchō-Kō ki*. (Tokyo: Jinbutsu Ōraisha), p. 259; Elisonas, J.S.A. & Lamers, J.P. (Trans. and Ed.) (2011) *The Chronicle of Lord Nobunaga by Ōta Gyūichi*. (Leiden: Brill), p. 324.
28. Hagiwara, T (Ed.) (1966). *Hōjō Godai ki* in *Sengoku Shiryō Sōsho Vol. 21*. (Tokyo: Jinbutsu Oraisha), p. 395.
29. Yamada, Yūji (2016). *Ninja no Rekishi*. (Tokyo, Kadokawa Shoten), p. 59.
30. Shimizu, Noboru (2009). *Sengoku ninja wa rekishi o dō ugokashita no ka?* (Tokyo, Best Shinsho); pp. 97-102.
31. Yoshida Y. (Ed.) (1979). *Taikō ki* Volume 1 (Tokyo, Kyōikusha), p. 185.
32. A printed version of it is included in a large and important collection of historical documents called *Gunsho ruijū* compiled early in the nineteenth century by the scholar Hanawa Koinichi (1746-1821). See Hanawa, Hoki(no)ichi (1960). *Gunsho Ruijū Volume 27* (kassen bu) (Tokyo, Zoku Gunsho Ruijū Kanseikai), p. 324.
33. Yoshida Y. (Ed.) (1979). *Taikō ki* Volume 1 (Tokyo, Kyōikusha), p. 185.
34. Iguchi, Asao (1995). *Maeda Toshiie* (Tokyo, Seibido Shuppan), p. 61. See also a related article in Various Authors (2017). *Ninja to yōkai* (Kwai no. 50) (Tokyo, Kadokawa Shoten), (Plates Section).
35. Shimizu, Noboru (2008). *Sengoku ninja retsuden* (Tokyo, Kawade Shobo Shinsha), pp. 186-190; Nakashima, Atsumi (2016) *Ninja o kagaku suru* (Tokyo, Yōsensha), pp. 44-45.
36. The text is in Sasama, Yoshihiko (1968). *Buke Senjin Sakuhō Shūsei*. (Tokyo, Yūzankaku), pp. 84.
37. Yamada, Yūji (2016). *Ninja no Rekishi*. (Tokyo, Kadokawa Shoten), p. 45.
38. This is the classic 'ninja' story about the flag-stealer of Hataya Castle which is

related below. The 1698 reference is found in Imamura, Y. (Ed.) (2005). *Ōu Eikei Gunki*. (Tokyo, Jinbutsu Oraisha), p. 185; *Mikawa Gofudo ki* of 1833 is in Kuwata, Tadachika & Utagawa, T. (Eds.) (1976). *Kaisei Mikawa Gofudo ki*. (Tokyo, Akita Shoten), Vol. 3 p. 442.

39. Sugiyama, Hiroshi (1974). *Sengoku Daimyo*. (Tokyo, Chūō Kōronsha). pp. 205-206.
40. Hobsbawm, E. & Ranger, T. (1983). (Eds.) *The Invention of Tradition*. (Cambridge University Press), p. 1.
41. Hobsbawm, E. & Ranger, T. (Eds.) (1983). *The Invention of Tradition*. (Cambridge University Press), pp. 4 & 7.
42. Hobsbawm, E. & Ranger, T. (Eds.) (1983). *The Invention of Tradition*. (Cambridge University Press), pp. 21-22.
43. For the surprising appearance of authentic pirate dress (which included bowler hats!) see Konstam, Angus (2011). *Pirate: The Golden Age*. (Oxford, Osprey Publishing).
44. Vlastos, Stephen (Ed.) (1998). *Mirror of Modernity: Invented Traditions of Modern Japan*. (Berkeley, University of California Press).
45. Smith, Robert J. 'Wedding and funeral ritual: analysing a moving target' in Van Bremen, Jan and Martinez, D.P. (eds.) (1995). *Ceremony and Ritual in Japan: Religious practices in an industrialized society*. (London, Routledge), p. 26.
46. Adolphson, Mikael & Commons, Anne (eds.) (2015). *Loveable Losers: The Heike in Action and Memory* (Honolulu, Univ of Hawaii Press).
47. Nitobe, Inazu (1905). *Bushido: The Soul of Japan: Tenth Revised Edition*. (New York: Putnam's). For a detailed and illuminating study of the complex emergence of *bushidō* see Benesch, Oleg (2014) *Inventing the Way of the Samurai: Nationalism, Internationalism, and Bushidō in Modern Japan* (Oxford, Oxford University Press).
48. The origins and extent of the Iga ninja tourist industry is discussed in a later chapter. In terms of local concentration and exaggeration of a historical theme a good parallel is the exploitation of the Dracula idea in certain areas of Romania.
49. Anonymous (1954). *Nagel's Travel Guide to Japan* (Geneva, Nagel Publishers), p. 544.
50. Yūki, S. (1988). 'Kōka jōkaku gun' *Rekishi Dokuhon Vol. 482* August 1988 pp. 114-121; Yokoyama, Takaharu (1992). *Nobunaga to Ise.Iga - Mie Sengoku monogatari* (Tokyo, Sōgensha), p. 29.
51. Yamamoto, Taketoshi (2016). *Nihon no intelligence kōsaku* (Tokyo; Shinyosha), pp. 3, 5 & 34.
52. Ferejohn, J.A. & Rosenbluth, F.M. (Eds.) (2010). *War and State Building in Medieval Japan*. (Stanford: Stanford University Press), p. 1.
53. Sugiyama, Hiroshi. (1974). *Sengoku Daimyo*. (Tokyo, Chūō Kōronsha), p. 205.

Chapter 2

1. Aston W.G. (1896). *'Nihongi: Chronicles of Japan from the Earliest times to A.D. 697 Volume II' Transactions and Proceedings of the Japan Society of London* Supplement I, p 125
2. Yoshimaru, Katsuya; Yamada, Yūji; Onishi, Yasumitsu *et. al.* (2014). *Ninja bungei kenkyū dokuhon* (Tokyo, Kasamashoin), p. 36.
3. Rabinovitch, Judith (1986).*Shōmonki, the story of Masakado's rebellion* (Tokyo, Monumenta Nipponica Monographs), pp. 92-95.

4. Shinoda, Minoru (1960). *The Founding of the Kamakura Shogunate 1180-1185* (New York, Columbia University), p. 157.
5. Yoshimaru, Katsuya; Yamada, Yūji; Onishi, Yasumitsu *et. al.* (2014). *Ninja bungei kenkyū dokuhon* (Tokyo, Kasamashoin), p. 36.
6. Griffith, Samuel B. (trans.) (1963). *Sun Tzu: The Art of War* (Oxford, Oxford University Press), pp. 144-149.
7. Varley, H. Paul (1994). *Warriors of Japan as portrayed in the war tales* (Honolulu, University of Hawaii Press), p. 42.
8. Diosy, Arthur (1911). 'Yoshitsune, the boy hero of Japan ' *Transactions and Proceedings of the Japan Society of London 10*, p. 60.
9. Shimizu, Noboru (2009). *Sengoku ninja wa rekishi o dō ugokashita no ka?* (Tokyo, Best Shinsho), p. 34.
10. Yoshimaru, Katsuya; Yamada, Yūji; Onishi, Yasumitsu *et. al.* (2014).*Ninja bungei kenkyū dokuhon* (Tokyo, Kasamashoin), pp. 40-41.
11. Sasama, Yoshihiko (1968). *Buke Senjin Sakuhō Shūsei.* (Tokyo, Yūzankaku), p. 83.
12. Owada, Tetsuo (2006). *Hideyoshi no tenka tōitsu sensō* (Tokyo, Yoshikawa Kobunkan), p. 129.
13. Imamura, Y. (Ed.) (2005). *Ōu Eikei Gunki.* (Tokyo, Jinbutsu Oraisha), p. 185.
14. This is the man who becomes a shinobi no mono in *Mikawa Gofudo ki* of 1833: Kuwata, Tadachika & Utagawa, T. (Eds.) (1976). *Kaisei Mikawa Gofudo ki.* (Tokyo, Akita Shoten), Vol. 3 p. 442.
15. Matsubara Kazuyoshi (2005). '*Mogami Yoshimitsu Monogatari* no kaisetsu to honkoku Part 2' *Bulletin of Naruto Kyoiku University Human and Social Sciences* Vol. 20, p. 5.
16. Shimizu, Noboru (2009). *Sengoku ninja wa rekishi o dō ugokashita no ka?* (Tokyo, Best Shinsho), p. 25-33.
17. Yamada, Yūji (2016). *Ninja no Rekishi.* (Tokyo, Kadokawa Shoten), pp. 26-27.
18. Fujiki, Hisashi (2005). *Zōhyōtachi no senjō* (Tokyo, Asahi Shinbun), pp. 130-131.
19. Yamada, Yūji (2016). *Ninja no Rekishi.* (Tokyo, Kadokawa Shoten), p. 45.
20. Sasama, Yoshihiko (1968). *Buke Senjin Sakuhō Shūsei.* (Tokyo, Yūzankaku), p. 380.
21. Yamada, Yūji (2016). *Ninja no Rekishi.* (Tokyo, Kadokawa Shoten), pp. 46-47.
22. Yamada, Yūji (2016). *Ninja no Rekishi.* (Tokyo, Kadokawa Shoten), p. 49.
23. I have translated *Mineaiki* as it appears in Yamada, Yūji (2016). *Ninja no Rekishi.* (Tokyo, Kadokawa Shoten), p. 28.
24. Two interesting articles about Medieval *akutō* are Oxenboell, Morten (2005).'Images of "Akutō"' *Monumenta Nipponica*, 60, 2, pp 235-262 and Oxenboell, Morten (2006). 'Mineaiki and Discourses on Social Unrest in Medieval Japan' *Japan Forum* 18, 1, pp. 1-21.
25. Friday, Karl (2004). *Samurai, warfare and the state in early medieval Japan* (New York, Routledge), p. 56.
26. Jansen, Marius B. (1995). *Warrior Rule in Japan* (New York, Cambridge University Press), p. 74.
27. Yamada, Yūji (2016). *Ninja no Rekishi.* (Tokyo, Kadokawa Shoten), p. 28.
28. Fujiki, Hisashi (2005). *Zōhyōtachi no senjō* (Tokyo, Asahi Shinbun), pp. 130-131; Yamada, Yūji (2016) *Ninja no Rekishi.* (Tokyo, Kadokawa Shoten), p. 50.
29. Elisonas, J.S.A. & Lamers, J.P. (Trans. and Ed.) (2011). *The Chronicle of Lord Nobunaga by Ōta Gyūichi.* (Leiden: Brill), p. 328,n.
30. Yamada, Yūji (2016). *Ninja no Rekishi.* (Tokyo, Kadokawa Shoten), p. 29.

31. Yamada, Yūji (2016). *Ninja no Rekishi*. (Tokyo, Kadokawa Shoten), pp. 30-32.
32. As translated by Oxenboell, Morten (2005). in 'Images of "Akutō"' *Monumenta Nipponica*, 60, 2, p. 245.
33. See for example Urban, William (2006).*Medieval Mercenaries: The Business of War*. (London, Greenhill Books),
34. Yamada, Yūji (2016). *Ninja no Rekishi*. (Tokyo, Kadokawa Shoten), p. 29
35. Fujiki, Hisashi (2005). *Zōhyōtachi no senjō: chūsei no yōhei to dorei kari* (Tokyo, Asahi Shinbun).
36. Morimoto, Masahiro (2014). *Kyōkai arasoi to Sengoku chōhōsen*. (Tokyo, Yōsensha), pp. 169-175.
37. Morimoto, Masahiro (2014). *Kyōkai arasoi to Sengoku chōhōsen*. (Tokyo, Yōsensha), p. 172.
38. Hagiwara, T (Ed.) (1966). *Hōjō Godai ki* in *Sengoku Shiryō Sōsho Vol. 21*. (Tokyo: Jinbutsu Oraisha), p. 395.
39. Hagiwara, T (Ed.) (1966). *Hōjō Godai ki* in *Sengoku Shiryō Sōsho Vol. 21*. (Tokyo: Jinbutsu Oraisha), pp. 397-398.
40. Hagiwara, T (Ed.)(1966). *Hōjō Godai ki* in *Sengoku Shiryō Sōsho Vol. 21*. (Tokyo: Jinbutsu Oraisha), pp. 397-398.
41. Fukurokuju is one of Japan's 'Seven gods of good fortune' and has an elongated head.
42. Hagiwara, T (Ed.) (1966). *Hōjō Godai ki* in *Sengoku Shiryō Sōsho Vol. 21*. (Tokyo: Jinbutsu Oraisha), p. 395.
43. Hagiwara, T (Ed.) (1966). *Hōjō Godai ki* in *Sengoku Shiryō Sōsho Vol. 21*. (Tokyo: Jinbutsu Oraisha), pp. 395-397.
44. Fujiki, Hisashi (2005). *Zōhyōtachi no senjō* (Tokyo, Asahi Shinbun), pp. 19-20.
45. Fujiki, Hisashi (2005). *Zōhyōtachi no senjō* (Tokyo, Asahi Shinbun), p. 136.
46. Morimoto, Masahiro (2014). *Kyōkai arasoi to Sengoku chōhōsen*. (Tokyo, Yōsensha), pp. 144-149.
47. Yamada, Yūji (2016). *Ninja no Rekishi*. (Tokyo, Kadokawa Shoten), p. 61.
48. Fujiki, Hisashi (2005). *Zōhyōtachi no senjō: chūsei no yōhei to dorei kari* (Tokyo, Asahi Shinbun), p. 33.
49. Fujiki, Hisashi (2005). *Zōhyōtachi no senjō: chūsei no yōhei to dorei kari* (Tokyo, Asahi Shinbun), pp. 27-28.
50. Fujiki, Hisashi (2005). *Zōhyōtachi no senjō: chūsei no yōhei to dorei kari* (Tokyo, Asahi Shinbun), p. 17.
51. Fujiki, Hisashi (2005). *Zōhyōtachi no senjō: chūsei no yōhei to dorei kari* (Tokyo, Asahi Shinbun), p. 19.
52. Kikuoka Jōgen (2006). *Iga Kyūkō Iran ki* (Iga-Ueno, Iga Kobunken Kankōkai), p. 27.
53. Kikuoka Jōgen (2006). *Iga Kyūkō Iran ki* (Iga-Ueno, Iga Kobunken Kankōkai), p. 27.
54. Kawakami, Jinichi (2016). *Ninja on okite* (Tokyo: Kadokawa Shoten), p. 19.
55. Various Authors (2016). *The Ninja - ninjatte nanja!? - kōshiki bukku* (Tokyo, Asahi Shimbun), pp. 10-11.

Chapter 3

1. Momochi, Orinosuke (Ed.) (1897). *Kōsei Iran ki* (Iga-Ueno), 7, 14-16.
2. Kuwata, Tadachika & Utagawa, T. (Eds.) (1976). *Kaisei Mikawa Gofudo ki*. (Tokyo,

Akita Shoten), Vol. 3 p. 197.
3. Nakashima, Atsumi (2016). *Ninja o kagaku suru* (Tokyo, Yōsensha). p. 48.
4. The name of the mountain is Kasagiyama; the alternative name given here refers to the presence of huge carvings of Maitreya (Miroku Bōsatsu) the Buddha of the Future on the rocks below the summit. They are still visible, although only the halo survives on the largest of them.
5. Takeuchi, Rizō (Ed.) (1978). *Tamon'In Nikki (Zōho Zoku Shiryō taisei Volume 38)* (Rinsen Shoten, Kyoto), p. 256.
6. Takeuchi, Rizō (Ed.) (1978). *Tamon'In Nikki (Zōho Zoku Shiryō taisei Volume 38)* (Rinsen Shoten, Kyoto), p. 258.
7. *Kyōroku Temmon no ki* is available in Nara Joshi Daigaku (1991). *Nara Joshi Daigaku kyōiku kenkyū nai tokubetsu keihi (Nara bunka ni kansuru sōgōteki kenkyū) hōkokusho.* (Nara, Nara Women's University), pp. 1-14. The Takada action appears on p. 7.
8. Nara Joshi Daigaku (1991). *Nara Joshi Daigaku kyōiku kenkyū nai tokubetsu keihi (Nara bunka ni kansuru sōgōteki kenkyū) hōkokusho.* (Nara, Nara Women's University), p. 10.
9. Kōyasan Shihensan Shohen. (1936) *Kōyasan Bunsho Volume 9* (Kōyasan Bunsho Kankō), Document 137, pp. 190-191. It may also be found in Okuno Takahiro (1988) *Oda Nobunaga monjo no kenkyū* (Tokyo, Yoshikawa Kōbunkan) pp. 544-546.
10. Kawakami, Jinichi (2016). *Ninja no okite* (Tokyo: Kadokawa Shoten), pp. 151-159.
11. Kawakami, Jinichi (2016). *Ninja no okite* (Tokyo: Kadokawa Shoten), pp. 151-159.
12. *Azai Sandai ki* Volume 10 http://yoshiok26.p1.bindsite.jp/bunken/cn14/asai10.html (Accessed 02 March 2016). Owada, Tetsuo (1973) *Ōmi Azai Shi* (Tokyo, Shin Jinbutsu Oraisha), p. 84.
13. Birt, Michael Patrick (1983). *Warring States: A study of the Go-Hōjō daimyo and domain 1491-1590* (Ph.D Thesis, Princeton University), p. 47.
14. Birt, Michael Patrick (1983). *Warring States: A study of the Go-Hōjō daimyo and domain 1491-1590* (Ph.D Thesis, Princeton University), p. 168.
15. Elisonas, Jurgis (1991). 'Christianity and the daimyo' In Hall, J. W. & McLain, J. L. (Eds.) *The Cambridge History of Japan. Vol. 4 Early modern Japan.* Cambridge: Cambridge University Press, p. 365.
16. Okuno Takahiro (1998). *Nobunaga monjo no kenkyū*, doc. 533.
17. Kuwata, Tadachika. (Ed.) (1965). *Shinchō-Kō ki.* (Tokyo: Jinbutsu Ōraisha), pp. 365-366; Elisonas, J.S.A. & Lamers, J.P. (Trans. and Ed.) (2011) *The Chronicle of Lord Nobunaga by Ōta Gyūichi.* (Leiden: Brill), p. 448.
18. Yoshimaru, Katsuya; Yamada, Yūji; Onishi, Yasumitsu *et. al.* (2014) *Ninja bungei kenkyū dokuhon* (Tokyo, Kasamashoin), p. 42; Iga City (2011) *Iga-shi shi Volume 1* (Iga City), p. 752.
19. Souyri, P. (2010). 'Autonomy and War in the Sixteenth Century Iga Region and the Birth of the Ninja Phenomena', in Ferejohn, J. A. & Rosenbluth, F. M. (Eds) *War and State Building in Medieval Japan.* (Stanford University Press), pp. 110-123.
20. Kurushima, Noriko (2011). *Ikki no sekai to hō* (Tokyo, Yamakawa), pp. 79-82.
21. Turnbull, Stephen (1991). *Ninja: The True Story of Japan's Secret Warrior Cult*, (London: Firebird Books), p.29.
22. Asawaka, K. (1929). *The Documents of Iriki: Illustrative of the Development of the Feudal Institutions of Japan.* (Yale University Press), p. 25.

23. Kuwata, Tadachika. (Ed.) (1965). *Shinchō-Kō ki*. (Tokyo: Jinbutsu Ōraisha), p. 105; Elisonas, J.S.A. & Lamers, J.P. (Trans. and Ed.) (2011) *The Chronicle of Lord Nobunaga by Ōta Gyūichi*. (Leiden: Brill), p. 145.

24. Kuwata, Tadachika. (Ed.) (1965). *Shinchō-Kō ki*. (Tokyo: Jinbutsu Ōraisha), p. 151; Elisonas, J.S.A. & Lamers, J.P. (Trans. and Ed.) (2011) *The Chronicle of Lord Nobunaga by Ōta Gyūichi*. (Leiden: Brill), p. 201.

25. Araki, Eishi (1987). *Higo Kunishū ikki* (Kumamoto: Kumamoto Shuppan Bunka Kaikan), p. 191.

26. Araki, Eishi (1987). *Higo Kunishū ikki* (Kumamoto: Kumamoto Shuppan Bunka Kaikan), pp. 92-93; Ōyama Ryūshū. (2003) *Hideyoshi to Higo Kunishū ikki* (Fukuoka, Kaichosha), pp. 124-126.

27. Kumamoto City (2000). *Kumamoto-shi shi kankei shiryōshū. Volume 4: Higo koki shūran*. (Kumamoto City), p. 431.

28. Kumamoto City (2000). *Kumamoto-shi shi kankei shiryōshū. Volume 4: Higo koki shūran*. (Kumamoto City), p. 65.

29. Kumamoto City (2000). *Kumamoto-shi shi kankei shiryōshū. Volume 4: Higo koki shūran*. (Kumamoto City), p. 66.

30. Tsuruta, Sōzō (1981).'Tenshō Amakusa Kassen no kōsatsu' *Kumamoto Shigaku* 55, pp. 49-50. Full a full account see Turnbull, Stephen (2013). 'The ghosts of Amakusa: localised opposition to centralised control in Higo Province, 1589-90.' *Japan Forum* 25 (2), pp. 191-211.

31. Araki, Eishi (1987). *Higo Kunishū ikki* (Kumamoto: Kumamoto Shuppan Bunka Kaikan), p. 138.

32. Tsuruta, Sōzō (1981). 'Tenshō Amakusa Kassen no kōsatsu' *Kumamoto Shigaku 55*, pp. 50 & 58.

Chapter 4

1. Yūki, S. (1988). 'Kōka jōkaku gun' *Rekishi Dokuhon Vol. 482 August 1988* pp. 114-121.

2. Fujita, Kazutoshi (2012). *'Kōka ninja' no jitsuzō*. (Tokyo, Yoshikawa Kōbunkan), pp. 13-14.

3. Okuno, Takahiro (1960). *Ashikaga Yoshiaki* (Tokyo, Yoshikawa Kōbunkan), p. 103; Yūki, S. (1988). *'Kōka jōkaku gun' Rekishi Dokuhon Vol. 482 August 1988* pp. 116-117. The castle sites are excellently presented using maps and an aerial photograph in a pamphlet available at the well-preserved and very interesting site. Kōka City Board of Education (2011). *Wada jōkan gun. (Kōka no jōkaku 1)* (Kōka City).

4. Fujita, Kazutoshi (2012). *'Kōka ninja' no jitsuzō*. (Tokyo, Yoshikawa Kōbunkan), p. 15.

5. Yūki, S. (1988). *'Kōka jōkaku gun' Rekishi Dokuhon Vol. 482 August 1988* p. 118.

6. Fujita, Kazutoshi (2012). *'Kōka ninja' no jitsuzō*. (Tokyo, Yoshikawa Kōbunkan), pp. 23-24.

7. That celebrated incident, much reproduced in Japanese art, was part of the great samurai tradition of being first into battle, although Takatsuna cheated by telling his opponent that his saddle girth was loose.

8. Shinya, Kazuyuki (2015). *Ōmi Rokkaku shi*. (Tokyo, Ebisu-Kosyo), p. 7.

9. Shimizu, Noboru (2008). *Sengoku ninja retsuden* (Tokyo, Kawade Shobo Shinsha), pp. 186-190; Nakashima, Atsumi (2016) *Ninja o kagaku suru* (Tokyo, Yōsensha), pp. 44-45.

10. Owada, Tetsuo (1973). *Ōmi Azai Shi* (Tokyo, Shin Jinbutsu Oraisha), pp. 352-353.
11. Lamers, J. P. (2000). *Japonius Tyrannus: The Japanese Warlord Oda Nobunaga Reconsidered*. (Leiden, Hotei Publishing), p. 42.
12. Kido, Masayuki (2008). 'O[da]Toyo[tomi]ki no Kōka: Kōka no yakiuchi wa nakatta' *Kiyō* (Shiga Bunkazai) 21, p. 37.
13. Shinya, Kazuyuki (2015). *Ōmi Rokkaku shi*. (Tokyo, Ebisu-Kosyo), p. 22.
14. Kuwata, Tadachika. (Ed.) (1965). *Shinchō-Kō ki*. (Tokyo: Jinbutsu Ōraisha), p. 83; Elisonas, J.S.A. & Lamers, J.P. (Trans. and Ed.) (2011) *The Chronicle of Lord Nobunaga by Ōta Gyūichi*. (Leiden, Brill), p. 117.
15. Kuwata, Tadachika. (Ed.) (1965). *Shinchō-Kō ki*. (Tokyo: Jinbutsu Ōraisha), p. 112; Elisonas, J.S.A. & Lamers, J.P. (Trans. and Ed.) (2011) *The Chronicle of Lord Nobunaga by Ōta Gyūichi*. (Leiden, Brill), p. 154.
16. Hatai, Hiromu (1975). *Shugo ryōgoku taisei no kenkyū : Rokkaku Shi ryōgoku ni miru Kinai kinkokuteki hatten no tokushitsu*. (Tokyo, Yoshikawa Kōbunkan), p. 311.
17. Kuwata, Tadachika. (Ed.) (1965). *Shinchō-Kō ki*. (Tokyo: Jinbutsu Ōraisha), p. 85; Elisonas, J.S.A. & Lamers, J.P. (Trans. and Ed.) (2011) *The Chronicle of Lord Nobunaga by Ōta Gyūichi*. (Leiden: Brill), pp. 118-120.
18. Lamers, J. P. (2000). *Japonius Tyrannus: The Japanese Warlord Oda Nobunaga Reconsidered*. (Leiden, Hotei Publishing), pp. 58-59.
19. Kuwata, Tadachika & Utagawa, T. (Eds.) (1976). *Kaisei Mikawa Gofudo ki*. Vol. 3 (Tokyo, Akita Shoten), p. 309.
20. Both are in modern Rittō City. Kuwata, Tadachika (Ed.) (1965). *Shinchō-Kō ki*. (Tokyo: Jinbutsu Ōraisha), pp. 103-104; Elisonas, J.S.A. & Lamers, J.P. (Trans. and Ed.) (2011). *The Chronicle of Lord Nobunaga by Ōta Gyūichi*. (Leiden: Brill), pp. 143-144.
21. In later accounts the battle of Yasugawa takes place after the siege of Chōkōji, an action famous for a defiant gesture on the part of the defending general. In *Buke Jiki* (1673) Shibata Katsuie smashes the water jars and leads his garrison in a sally that carries all before them. The Rokkaku force are then pursued to Yasugawa, rather than raiding there as in *Shinchō-Kō ki*.
22. Ochikubo is now a district of Yasu City. Kuwata, Tadachika (Ed.) (1965). *Shinchō-Kō ki*. (Tokyo: Jinbutsu Ōraisha), pp. 105; Elisonas, J.S.A. & Lamers, J.P. (Trans. and Ed.) (2011). *The Chronicle of Lord Nobunaga by Ōta Gyūichi*. (Leiden: Brill), p. 145.
23. Kuwata, Tadachika (Ed.) (1965). *Shinchō-Kō ki*. (Tokyo: Jinbutsu Ōraisha), pp. 114; Elisonas, J.S.A. & Lamers, J.P. (Trans. and Ed.) (2011). *The Chronicle of Lord Nobunaga by Ōta Gyūichi*. (Leiden: Brill), p. 157.
24. Kuwata, Tadachika (Ed.) (1965). *Shinchō-Kō ki*. (Tokyo: Jinbutsu Ōraisha), p. 104; Elisonas, J.S.A. & Lamers, J.P. (Trans. and Ed.) (2011). *The Chronicle of Lord Nobunaga by Ōta Gyūichi*. (Leiden: Brill), pp. 144-145.
25. Kuwata, Tadachika (Ed.) (1965). *Shinchō-Kō ki*. (Tokyo: Jinbutsu Ōraisha), p. 150; Elisonas, J.S.A. & Lamers, J.P. (Trans. and Ed.) (2011). *The Chronicle of Lord Nobunaga by Ōta Gyūichi*. (Leiden: Brill), pp. 199-200.
26. Kuwata, Tadachika (Ed.) (1965). *Shinchō-Kō ki*. (Tokyo: Jinbutsu Ōraisha), pp. 114-115; Elisonas, J.S.A. & Lamers, J.P. (Trans. and Ed.) (2011). *The Chronicle of Lord Nobunaga by Ōta Gyūichi*. (Leiden: Brill), pp. 158-159.
27. Kuwata, Tadachika (Ed.) (1965). *Shinchō-Kō ki*. (Tokyo: Jinbutsu Ōraisha), p. 136; Elisonas, J.S.A. & Lamers, J.P. (Trans. and Ed.) (2011). *The Chronicle of Lord*

Nobunaga by Ōta Gyūichi. (Leiden: Brill), p. 183.

28. Kuwata, Tadachika (Ed.) (1965). *Shinchō-Kō ki.* (Tokyo: Jinbutsu Ōraisha), p. 151; Elisonas, J.S.A. & Lamers, J.P. (Trans. and Ed.) (2011). *The Chronicle of Lord Nobunaga by Ōta Gyūichi.* (Leiden: Brill), p. 201.
29. Okuno, Takahiro (1960). *Ashikaga Yoshiaki* (Tokyo, Yoshikawa Kōbunkan), pp. 213-214.
30. Okuno, Takahiro (1960). *Ashikaga Yoshiaki* (Tokyo, Yoshikawa Kōbunkan), pp. 216-217 (which includes a map of the Uji area); Lamers, J. P. (2000). *Japonius Tyrannus: The Japanese Warlord Oda Nobunaga Reconsidered.* (Leiden, Hotei Publishing), pp. 96-97.
31. Fujita, Kazutoshi (2012). *'Kōka ninja' no jitsuzō.* (Tokyo, Yoshikawa Kōbunkan), pp. 24-25; Kido, Masayuki. (2008). 'O[da]Toyo[tomi]ki no Kōka: Kōka no yakiuchi wa nakatta' *Kiyō* (Shiga Bunkazai) 21, p. 39.
32. Kuwata, Tadachika (Ed.) (1965). *Shinchō-Kō ki.* (Tokyo: Jinbutsu Ōraisha), p. 156; Elisonas, J.S.A. & Lamers, J.P. (Trans. and Ed.) (2011) *The Chronicle of Lord Nobunaga by Ōta Gyūichi.* (Leiden: Brill), p. 207.
33. Shinya, Kazuyuki (2015). *Ōmi Rokkaku shi.* (Tokyo, Ebisu-Kosho), p. 23.
34. Kido, Masayuki (2008). 'O[da]Toyo[tomi]ki no Kōka: Kōka no yakiuchi wa nakatta' *Kiyō* (Shiga Bunkazai) 21, p. 40.
35. Kido, Masayuki (2008). 'O[da]Toyo[tomi]ki no Kōka: Kōka no yakiuchi wa nakatta' *Kiyō* (Shiga Bunkazai) 21, p. 43.
36. Kuwata, Tadachika (Ed.) (1965). *Shinchō-Kō ki.* (Tokyo: Jinbutsu Ōraisha), p. 332; Elisonas, J.S.A. & Lamers, J.P. (Trans. and Ed.) (2011) *The Chronicle of Lord Nobunaga by Ōta Gyūichi.* (Leiden: Brill), p. 410.
37. Kido, Masayuki (2008). 'O[da]Toyo[tomi]ki no Kōka: Kōka no yakiuchi wa nakatta' *Kiyō* (Shiga Bunkazai) 21, p. 40. For a detailed account of the Sōkenji and its removal from Kōka see Akita, Hiroki (1990) *Oda Nobunaga to Azuchijō* (Osaka, Sōgensha).
38. Kuwata, Tadachika (Ed.) (1965). *Shinchō-Kō ki.* (Tokyo: Jinbutsu Ōraisha), p. 343; Elisonas, J.S.A. & Lamers, J.P. (Trans. and Ed.) (2011) *The Chronicle of Lord Nobunaga by Ōta Gyūichi.* (Leiden: Brill), p. 421-423.

Chapter 5
1. Kuwata, Tadachika (Ed.) (1965). *Shinchō-Kō ki.* (Tokyo: Jinbutsu Ōraisha), pp. 262-163 & 332-335; Elisonas, J.S.A. & Lamers, J.P. (Trans. and Ed.) (2011) *The Chronicle of Lord Nobunaga by Ōta Gyūichi.* (Leiden: Brill), pp. 328-329 & 410-414.
2. Momochi, Orinosuke (Ed.) (1897). *Kōsei Iran ki* (Iga-Ueno). The text is available online from the National Diet Library at http://kindai.ndl.go.jp/info:ndljp/pid/768755 (Accessed 9 January 2016).
3. Kikuoka Jōgen (2006). *Iga Kyūkō Iran ki* (Iga-Ueno, Iga Kobunken Kankōkai).
4. Kikuoka Jōgen (2006). *Iga Kyūkō Iran ki* (Iga-Ueno, Iga Kobunken Kankōkai), p. 27.
5. Momochi, Orinosuke (Ed.) (1897). *Kōsei Iran ki* (Iga-Ueno), 4, 5.
6. Momochi, Orinosuke (Ed.) (1897). *Kōsei Iran ki* (Iga-Ueno), 1,7.
7. Another example of *kanchō* occurs in section 2,2.
8. Momochi, Orinosuke (Ed.) (1897). *Kōsei Iran ki* (Iga-Ueno), 4, 5.
9. For examples see Momochi, Orinosuke (Ed.) (1897). *Kōsei Iran ki* (Iga-Ueno), 6,5 and 3,13.
10. Momochi, Orinosuke (Ed.) (1897). *Kōsei Iran ki* (Iga-Ueno), 7, 14-16.

11. Momochi, Orinosuke (Ed.) (1897). *Kōsei Iran ki* (Iga-Ueno), 2,4.
12. Momochi, Orinosuke (Ed.) (1897). *Kōsei Iran ki* (Iga-Ueno), 2,1.
13. Lamers, J. P. (2000). *Japonius Tyrannus: The Japanese Warlord Oda Nobunaga Reconsidered.* (Leiden, Hotei Publishing), p. 68.
14. Yokoyama, Takaharu (1992). *Nobunaga to Ise.Iga - Mie Sengoku monogatari* (Tokyo, Sōgensha), p. 119.
15. Momochi, Orinosuke (Ed.) (1897). *Kōsei Iran ki* (Iga-Ueno), 2,1.
16. Momochi, Orinosuke (Ed.) (1897). *Kōsei Iran ki* (Iga-Ueno), 2,1.
17. Momochi, Orinosuke (Ed.) (1897). *Kōsei Iran ki* (Iga-Ueno), 2,3.
18. Momochi, Orinosuke (Ed.) (1897). *Kōsei Iran ki* (Iga-Ueno), 2,4.
19. Momochi, Orinosuke (Ed.) (1897). *Kōsei Iran ki* (Iga-Ueno), 2,4.
20. Momochi, Orinosuke (Ed.) (1897). *Kōsei Iran ki* (Iga-Ueno), 2,6.
21. The identifying flags worn on the back of a suit of armour.
22. Momochi, Orinosuke (Ed.) (1897). *Kōsei Iran ki* (Iga-Ueno), 2,8.
23. Momochi, Orinosuke (Ed.) (1897). *Kōsei Iran ki* (Iga-Ueno), 2,8.
24. Momochi, Orinosuke (Ed.) (1897). *Kōsei Iran ki* (Iga-Ueno), 2,11.
25. Kuwata, Tadachika (Ed.) (1965). *Shinchō-Kō ki.* (Tokyo: Jinbutsu Ōraisha), p. 262; Elisonas, J.S.A. & Lamers, J.P. (Trans. and Ed.) (2011). *The Chronicle of Lord Nobunaga by Ōta Gyūichi.* (Leiden: Brill), p. 328.
26. Kuwata, Tadachika (Ed.) (1965). *Shinchō-Kō ki.* (Tokyo: Jinbutsu Ōraisha), pp. 262-263; Elisonas, J.S.A. & Lamers, J.P. (Trans. and Ed.) (2011). *The Chronicle of Lord Nobunaga by Ōta Gyūichi.* (Leiden: Brill), p. 329.
27. Kuwata, Tadachika (Ed.) (1965). *Shinchō-Kō ki.* (Tokyo: Jinbutsu Ōraisha), p. 332; Elisonas, J.S.A. & Lamers, J.P. (Trans. and Ed.) (2011). *The Chronicle of Lord Nobunaga by Ōta Gyūichi.* (Leiden: Brill), p. 410.
28. Kuwata, Tadachika (Ed.) (1965). *Shinchō-Kō ki.* (Tokyo: Jinbutsu Ōraisha), p. 332; Elisonas, J.S.A. & Lamers, J.P. (Trans. and Ed.) (2011). *The Chronicle of Lord Nobunaga by Ōta Gyūichi.* (Leiden: Brill), p. 410.
29. Momochi, Orinosuke (Ed.) (1897). *Kōsei Iran ki* (Iga-Ueno), 3,5.
30. Momochi, Orinosuke (Ed.) (1897). *Kōsei Iran ki* (Iga-Ueno), 3,13.
31. Kuwata, Tadachika (Ed.) (1965). *Shinchō-Kō ki.* (Tokyo: Jinbutsu Ōraisha), p. 333; Elisonas, J.S.A. & Lamers, J.P. (Trans. and Ed.) (2011). *The Chronicle of Lord Nobunaga by Ōta Gyūichi.* (Leiden: Brill), p. 411.
32. Momochi, Orinosuke (Ed.) (1897). *Kōsei Iran ki* (Iga-Ueno), 3,14.
33. Takeuchi, Rizō (Ed.) (1978). *Tamon'In Nikki (Zōho Zoku Shiryō taisei Volume 39)* (Rinsen Shoten, Kyoto), p. 240.
34. Momochi, Orinosuke (Ed.) (1897). *Kōsei Iran ki* (Iga-Ueno), 3,14.
35. Momochi, Orinosuke (Ed.) (1897). *Kōsei Iran ki* (Iga-Ueno), 4,9.
36. Kikuoka Jōgen (2006). *Iga Kyūkō Iran ki* (Iga-Ueno, Iga Kobunken Kankōkai), p. 59.
37. Momochi, Orinosuke (Ed.) (1897). *Kōsei Iran ki* (Iga-Ueno), 5,11.
38. Momochi, Orinosuke (Ed.) (1897). *Kōsei Iran ki* (Iga-Ueno), 6,1.
39. Momochi, Orinosuke (Ed.) (1897). *Kōsei Iran ki* (Iga-Ueno), 6,2.
40. Momochi, Orinosuke (Ed.) (1897). *Kōsei Iran ki* (Iga-Ueno), 6,7.
41. Momochi, Orinosuke (Ed.) (1897). *Kōsei Iran ki* (Iga-Ueno), 6,7.
42. Kuwata, Tadachika (Ed.) (1965). *Shinchō-Kō ki.* (Tokyo: Jinbutsu Ōraisha), p. 334; Elisonas, J.S.A. & Lamers, J.P. (Trans. and Ed.) (2011). *The Chronicle of Lord Nobunaga by Ōta Gyūichi.* (Leiden: Brill), p. 412.

43. Momochi, Orinosuke (Ed.) (1897). *Kōsei Iran ki* (Iga-Ueno), 7,1.
44. Momochi, Orinosuke (Ed.) (1897). *Kōsei Iran ki* (Iga-Ueno), 7,3.
45. Momochi, Orinosuke (Ed.) (1897). *Kōsei Iran ki* (Iga-Ueno), 7,2.
46. Momochi, Orinosuke (Ed.) (1897). *Kōsei Iran ki* (Iga-Ueno), 7, 14-16.
47. Momochi, Orinosuke (Ed.) (1897). *Kōsei Iran ki* (Iga-Ueno), 7, 16-17.
48. Kuwata, Tadachika (Ed.) (1965). *Shinchō-Kō ki*. (Tokyo: Jinbutsu Ōraisha), pp. 335-336; Elisonas, J.S.A. & Lamers, J.P. (Trans. and Ed.) (2011). *The Chronicle of Lord Nobunaga by Ōta Gyūichi*. (Leiden: Brill), p. 414.
49. Momochi, Orinosuke (Ed.) (1897). *Kōsei Iran ki* (Iga-Ueno), 4, 5.
50. Yokoyama, Takaharu (1992). *Nobunaga to Ise.Iga - Mie Sengoku monogatari* (Tokyo, Sōgensha), p. 153.
51. Kuwata, Tadachika (Ed.) (1965). *Shinchō-Kō ki*. (Tokyo: Jinbutsu Ōraisha), p. 334; Elisonas, J.S.A. & Lamers, J.P. (Trans. and Ed.) (2011). *The Chronicle of Lord Nobunaga by Ōta Gyūichi*. (Leiden: Brill), p. 413.
52. Yamada, Yūji (2014). *Ninja no Kyōkasho: Shin Mansenshūkai*. (Tokyo: Kasamashoin), p. 13.
53. Kuwata, Tadachika & Utagawa, T. (Eds.) (1976). *Kaisei Mikawa Gofudo ki*. (Tokyo: Akita Shoten), Vol. 2, p. 191. A brief account is in Yoshida Y. (Ed.) (1979). *Taikō ki* Volume 1 (Tokyo, Kyōikusha), p. 146.
54. Yamada, Yūji (2014). *Ninja no Kyōkasho: Shin Mansenshūkai*. (Tokyo: Kasamashoin), p. 13.
55. See for example Souyri, P. (2010). 'Autonomy and War in the Sixteenth Century Iga Region and the Birth of the Ninja Phenomena', in Ferejohn, J. A. & Rosenbluth, F. M. (Eds) (2010) *War and State Building in Medieval Japan*. (Stanford University Press), p. 122.
56. Yoshida Y. (Ed.) (1979). *Taikō ki* Volume 3 (Tokyo, Kyōikusha), pp. 107-108.
57. Park, Y. (1978). *Admiral Yi and his Turtleboat Armada*. (Seoul: Hanjin Publishing Company), p. 104.
58. Lamers, J. P. (2000). *Japonius Tyrannus: The Japanese Warlord Oda Nobunaga Reconsidered*. (Leiden, Hotei Publishing), p. 204.
59. Owada, Tetsuo (2006). *Hideyoshi no tenka tōitsu sensō* (Tokyo, Yoshikawa Kobunkan), p. 152.
60. Owada, Tetsuo (2006). *Hideyoshi no tenka tōitsu sensō* (Tokyo, Yoshikawa Kobunkan), p. 154.
61. Yoshida Y. (Ed.) (1979). *Taikō ki* Volume 2 (Tokyo, Kyōikusha), pp. 56-59.
62. Kuwata, Tadachika (1975). *Toyotomi Hideyoshi Kenkyū*. (Tokyo, Kadokawa Shoten), p. 193.
63. Anotsu is the modern city of Tsu, the prefectural capital of Mie.
64. Anonymous (1891) *Tourist's Guide and Interpreter* (Tokyo, Seishi Bunsha), p. 65.

Chapter 6

1. Owada, Tetsuo (2006). *Hideyoshi no tenka tōitsu sensō* (Tokyo, Yoshikawa Kobunkan), p. 162-163.
2. *Dai Nihon Shiryō* (1975) (Tokyo, Tokyo University) Volume 11, Part 15, p. 60.
3. *Dai Nihon Shiryō* (1975) (Tokyo, Tokyo University) Volume 11, Part 15, p. 61.
4. *Dai Nihon Shiryō* (1975) (Tokyo, Tokyo University) Volume 11, Part 15, pp. 83 & 95.
5. *Dai Nihon Shiryō* (1975) (Tokyo, Tokyo University) Volume 11, Part 15, p. 95.
6. *Dai Nihon Shiryō* (1975) (Tokyo, Tokyo University) Volume 11, Part 15, p. 61.

7. Owada, Tetsuo (2006). *Hideyoshi no tenka tōitsu sensō* (Tokyo, Yoshikawa Kobunkan), p. 163.
8. Yoshida Y. (Ed.) (1979). *Taikō ki* Volume 2 (Tokyo, Kyōikusha), pp. 60-61.
9. Kido, Masayuki (2008). 'O[da]Toyo[tomi]ki no Kōka: Kōka no yakiuchi wa nakatta' *Kiyō* (Shiga Bunkazai) 21, p, 43; Fujita, Kazutoshi (2012). *'Kōka ninja' no jitsuzō*. (Tokyo, Yoshikawa Kōbunkan), p. 25.
10. *Dai Nihon Shiryō* (1975) (Tokyo, Tokyo University) Volume 11, Part 15, pp. 61, 83 & 95.
11. Kido, Masayuki (2008). 'O[da]Toyo[tomi]ki no Kōka: Kōka no yakiuchi wa nakatta' *Kiyō* (Shiga Bunkazai) 21, p. 43.
12. *Dai Nihon Shiryō* (1975) (Tokyo, Tokyo University) Volume 11, Part 15,,pp. 234-235).
13. Minakuchi Okayama Castle refers to the *yamashiro* built by Nakamura Kazuuji in 1585. The later Minakuchi Castle, part of which has been rebuilt as a museum, lies within the town and was built in 1634.
14. Fujita, Kazutoshi (2012). *'Kōka ninja' no jitsuzō*. (Tokyo, Yoshikawa Kōbunkan), p. 25.
15. Fujita, Kazutoshi (2012). *'Kōka ninja' no jitsuzō*. (Tokyo, Yoshikawa Kōbunkan), pp. 26-27. The shrine is referred to here as the Yagawa temple because it was then in the situation of the blending of Buddhism with Shintō
16. Fujiki, H. (2005). *Katanagari*. (Tokyo: Iwanami Shoten), p. 59.
17. Yamada, Yūji (2014). *Ninja no Kyōkasho: Shin Mansenshūkai*. (Tokyo, Kasamashoin), p. 16.
18. Dōami was the fourth son of the keeper of Seta Castle in Ōmi and as a child had been sent off to the monastery of Miidera, hence his Buddhist name.
19. Kuwata, Tadachika & Utagawa, T. (Eds.) (1976). *Kaisei Mikawa Gofudo ki*. (Tokyo, Akita Shoten), Vol. 3 p. 197.
20. Kuwata, Tadachika & Utagawa, T. (Eds.) (1976). *Kaisei Mikawa Gofudo ki*. (Tokyo, Akita Shoten), Vol. 3 p. 187, where the two unfortunate men are named as Yamaguchi Sōsuke and Eibara Jūnai.
21. Kuwata, Tadachika & Utagawa, T. (Eds.) (1976). *Kaisei Mikawa Gofudo ki*. (Tokyo, Akita Shoten), Vol. 3 p. 160.
22. The *koku* (278.3 litres) was a measure of a landowner's wealth based on the assessed yield from rice fields.
23. Fujita, Kazutoshi (2012). *'Kōka ninja' no jitsuzō*. (Tokyo, Yoshikawa Kōbunkan), p. 31.
24. Fujita, Kazutoshi (2012). *'Kōka ninja' no jitsuzō*. (Tokyo, Yoshikawa Kōbunkan), p. 53.
25. Gongen-sama refers to Tokugawa Ieyasu in his posthumous manifestation as the *gongen* (avatar) of the 'Sun God of the East, by which name he is enshrined at the Tōshōgū in Nikkō.
26. Fujita, Kazutoshi (2012). *'Kōka ninja' no jitsuzō*. (Tokyo, Yoshikawa Kōbunkan), pp. 35-36.
27. Kuwata, Tadachika & Utagawa, T. (Eds.) (1976). *Kaisei Mikawa Gofudo ki*. (Tokyo, Akita Shoten), p. 244.
28. Kuwata, Tadachika & Utagawa, T. (Eds.) (1976). *Kaisei Mikawa Gofudo ki*. (Tokyo, Akita Shoten), p. 262.
29. Fujita, Kazutoshi (2012). *'Kōka ninja' no jitsuzō*. (Tokyo, Yoshikawa Kōbunkan), p. 37.

30. Fujita, Kazutoshi (2012). *'Kōka ninja' no jitsuzō*. (Tokyo, Yoshikawa Kōbunkan), pp. 37-40.
31. Fujita, Kazutoshi (2012). *'Kōka ninja' no jitsuzō*. (Tokyo, Yoshikawa Kōbunkan), pp. 48-51.
32. Toda, Toshio (1988). *Amakusa-Shimabara no ran: Hosokawa han shiryō ni yoru.* (Tokyo: Shin Jinbutsu Ōraisha), pp. 198-199.
33. Toda, Toshio (1988). *Amakusa-Shimabara no ran: Hosokawa han shiryō ni yoru.* (Tokyo: Shin Jinbutsu Ōraisha), p. 191.
34. Toda, Toshio (1988). *Amakusa-Shimabara no ran: Hosokawa han shiryō ni yoru.* (Tokyo: Shin Jinbutsu Ōraisha), p. 199.
35. Okada, Akio (1960). *Amakusa Tokisada.* (Tokyo, Yoshikawa Kōbunkan), pp. 254-255.
36. Fujita, Kazutoshi (2012). *'Kōka ninja' no jitsuzō*. (Tokyo, Yoshikawa Kōbunkan), pp. 49-52.

Chapter 7
1. Takao, Yoshiki (2017). *Ninja no Matsue: Edo-jō ni tsutometa Iga-mono tachi* (Tokyo, Kadokawa), p. 18.
2. Fukai, Masaumi (1992). *Edojō oniwaban : Tokugawa Shōgun no mimi to me* (Tokyo, Chūō Kōronsha), p. 4.
3. Norman, E. Herbert (1940). *Japan's Emergence as a Modern State: Political and Economic Problems of the Meiji Period* (New York, Institute of Pacific Relations), p. 14.
4. Shimizu, Noboru (2008). *Sengoku ninja retsuden* (Tokyo, Kawade Shobo Shinsha), pp. 234-236.
5. Shimizu, Noboru (2009). *Edo no onmitsu: oniwaban* (Tokyo, Kawade Shobo Shinsha), p. 18.
6. Fukai, Masaumi (1992). *Edojō oniwaban : Tokugawa Shōgun no mimi to me* (Tokyo, Chūō Kōronsha), p. 2.
7. Nakashima, Atsumi (2016). *Ninja o kagaku suru* (Tokyo, Yōsensha), pp. 36-38.
8. Marcure, Kenneth (1985). 'The Danka System' *Monumenta Nipponica* 40, 1, p.43.
9. Shimizu, Noboru (2008). *Sengoku ninja retsuden* (Tokyo, Kawade Shobo Shinsha), p. 237.
10. Iwao, Seiichi (1934). 'Matsukura Shigemasa no Ruzonto ensei keikaku' *Shigaku Zasshi* 45, 9, p. 98.
11. Hayashi, Senkichi (ed.) (1973). *Shimabara Hantō-shi Volume II* (Nagasaki, Minami Takaki-gun Board of Education), p. 980; Nagasaki Prefecture *Nagasaki-ken shi*, p. 246.
12. Blair, Emma Helen & Robertson, James Alexander (Eds.) (1903-9). *The Philippine Islands, 1493-1803: (Volume 1 Part 24.).* (Cleveland, Arthur H. Clark Co.), pp. 245-246. For a full account of the operations see Turnbull, Stephen (2016). 'Wars and Rumours of Wars: Japanese Plans to invade the Philippines, 1593-1637' *Naval War College Review* Vol. 69, no, 4, pp. 107-120.
13. Schwade, Arcadio (1964). 'Matsukura Shigemasa no Ruzonto ensei keikaku' *Kirishitan Kenkyū* 9, p. 345.
14. Iwao, Seiichi (1934). 'Matsukura Shigemasa no Ruzonto ensei keikaku' *Shigaku Zasshi* 45, 9, p. 101.
15. Hayashi, Senkichi (ed.) (1954). *Shimabara Hantō-shi Volume II* (Nagasaki, Minami

Takaki-gun Board of Education), p. 980. The theory about his assassination is discussed in detail in Schwade, Arcadio (1964), pp. 346-348.
16. Sansom, George (1963). *A History of Japan: 1615-1867* (Stanford University Press), pp. 54-56
17. The text of the regulations is in Sasama, Yoshihiko (1968). *Buke Senjin Sakuhō Shūsei*. (Tokyo, Yūzankaku), pp. 81.
18. Yamada, Yūji (2016) *Ninja no Rekishi*. (Tokyo, Kadokawa Shoten), p. 214.
19. Nakashima, Atsumi (2016). *Ninja o kagaku suru* (Tokyo, Yōsensha), p. 36.
20. Roberts, Luke (1997). 'A Petition for a Popularly Chosen Council of Government in Tosa in 1787'. *Harvard Journal of Asiatic Studies* 57, 2 p. 575.
21. Roberts, Luke (1997). 'A Petition for a Popularly Chosen Council of Government in Tosa in 1787'. *Harvard Journal of Asiatic Studies* 57, 2 pp. 580-581.
22. For accounts of the Oniwaban see Fukai, Masaumi (1992). *Edojō oniwaban : Tokugawa Shōgun no mimi to me* (Tokyo, Chūō Kōronsha) and Shimizu, Noboru (2009). *Edo no onmitsu: oniwaban* (Tokyo, Kawade Shobo Shinsha).
23. For a list of these investigations between 1787 and 1839 see Fukai, Masaumi (1992). *Edojō oniwaban : Tokugawa Shōgun no mimi to me* (Tokyo, Chūō Kōronsha), pp. 69-71.
24. Shimizu, Noboru (2009). *Edo no onmitsu: oniwaban* (Tokyo, Kawade Shobo Shinsha), pp. 201-205.
25. Shimizu, Noboru (2009). *Edo no onmitsu: oniwaban* (Tokyo, Kawade Shobo Shinsha), p. 162.
26. From a collection of Nobeoka documents available online at http://bakumatu0982.jp/29report/report29-gunrei-150327.html (Accessed 09 March 2016).

Chapter 8
1. Morinaga, Maki Isaka (2005). *Secrecy in Japanese Arts: "Secret Transmission" as a Mode of Knowledge* (New York, Palgrave Maacmillan), pp. 1-2.
2. Various Authors (2002). *Iga.Kōka shinobi no subete* Bessatsu Rekishi Dokuhon 619. (Tokyo, Shinjinbutsu), Front cover.
3. Hori, Ichiro (1966). 'Mountains and their importance for the idea of the other world in Japanese folk religion' *History of Religions* 6, pp. 7-8.
4. Anonymous (1966). 'Editorial' *Black Belt*, 4, 12 December 1966, p. 11.
5. Sasama, Yoshihiko (1968). *Buke Senjin Sakuhō Shūsei*. (Tokyo, Yūzankaku), pp. 84.
6. Sasama, Yoshihiko (1968). *Buke Senjin Sakuhō Shūsei*. (Tokyo, Yūzankaku), p. 85.
7. Yamada, Yūji (2016). *Ninja no Rekishi*. (Tokyo, Kadokawa Shoten), p. 45.
8. Kikuoka Jōgen (2006). *Iga Kyūkō Iran ki* (Iga-Ueno, Iga Kobunken Kankōkai), p. 27.
9. The best commentary is provided in Nakashima, Atsumi (2017). *Ninja no heihō* (Tokyo, Kadokawa Shoten).
10. The most important manuals can be found in Cummins, Antony and Minami, Yoshiie (2011). *True Path of the Ninja* (Rutland VT, Tuttle); Cummins, Antony and Minami, Yoshiie (2012).*The Secret Traditions of the Shinobi* (Berkeley, Blue Snake Books); Cummins, Antony and Minami, Yoshiie (2013a). *The Book of Ninja* (London, Watkins Press); Cummins, Antony and Minami, Yoshiie (2013b). *Iga and Koga Ninja Skills* (Stroud, The History Press).

11. Yoshimaru, Katsuya; Yamada, Yūji; Onishi, Yasumitsu *et. al.* (2014). *Ninja bungei kenkyū dokuhon* (Tokyo, Kasamashoin), p. 39. An English language summary is available in Yamada, Yūji (2014). *The Spirit of Ninja: A Study of the Global Ninja Craze* (Tsu City, Mie University) E-book, loc 46.

12. The text is in Sasama, Yoshihiko (1968). *Buke Senjin Sakuhō Shūsei.* (Tokyo, Yūzankaku), pp. 380-403. It is discussed in Yoshimaru, Katsuya; Yamada, Yūji; Onishi, Yasumitsu *et. al.* (2014). *Ninja bungei kenkyū dokuhon* (Tokyo, Kasamashoin), p. 40.

13. Sasama, Yoshihiko (1968). *Buke Senjin Sakuhō Shūsei.* (Tokyo, Yūzankaku), p. 380.

14. Sasama, Yoshihiko (1968). *Buke Senjin Sakuhō Shūsei.* (Tokyo, Yūzankaku), p. 383-391.

15. Sasama, Yoshihiko (1968). *Buke Senjin Sakuhō Shūsei.* (Tokyo, Yūzankaku), pp. 392-393.

16. Yoshimaru, Katsuya; Yamada, Yūji; Onishi, Yasumitsu *et. al.* (2014). *Ninja bungei kenkyū dokuhon* (Tokyo, Kasamashoin), pp. 43-44.

17. Until recently the work was invariably known as *Bansenshūkai.* As none of the source copies give any indication how the title should be pronounced no definitive statement can be made.

18. The critical edition is Nakashima, Atsumi (2015). *Mansenshūkai* (Tokyo, Kokusho). It also is reproduced in its entirety in Imamura, Yoshio (Ed.) (1966). *Nihon budō zenshū Volume 4* (Tokyo, Jinbutsu Oraisha). For an English translation see Cummins, Antony and Minami, Yoshiie (2013a). *The Book of Ninja* (London, Watkins Press).

19. The expression even made it into the pages of the first popular book about ninja published by Mie University: Yamada, Yūji (2014). *Ninja no Kyōkasho: Shin Mansenshūkai.* (Tokyo, Kasamashoin), p. 28.

20. Imamura, Yoshio (Ed.) (1966). *Nihon budō zenshū Volume 4* (Tokyo, Jinbutsu Oraisha), p. 414.

21. Yamada, Fūtarō (2012). *Kunoichi Ninpōchō* (Tokyo, Kadokawa Shoten), p. 315.

22. Imamura, Yoshio (Ed.) (1966). *Nihon budō zenshū Volume 4* (Tokyo, Jinbutsu Oraisha), p. 428.

23. Hikone City Board of Education (no date). *'Watashi no machi no Sengoku': Dodo uji to Dodo yakata* (information sheet) (Hikone City).

24. Owada, Tetsuo (1973). *Ōmi Azai Shi* (Tokyo, Shin Jinbutsu Oraisha), p. 81.

25. Owada, Tetsuo (1973). *Ōmi Azai Shi* (Tokyo, Shin Jinbutsu Oraisha), pp. 82-84.

26. *Azai Sandai ki Volume 11* http://yoshiok26.p1.bindsite.jp/bunken/cn14/asai12.html (Accessed 03 March 2016); Hikone City Board of Education (no date). *'Watashi no machi no Sengoku': Dodo uji to Dodo yakata* (information sheet) (Hikone City).

27. Kikuoka Jōgen (2006). *Iga Kyūkō Iran ki* (Iga-Ueno, Iga Kobunken Kankōkai), p. 27.

28. Yamada, Yūji (2014). *Ninja no Kyōkasho: Shin Mansenshūkai.* (Tokyo, Kasamashoin), p. 29

29. Yamada, Yūji (2014). *Ninja no Kyōkasho: Shin Mansenshūkai.* (Tokyo, Kasamashoin), p. 29. A good idea of its contents may be gained from the augmented translation: Cummins, Antony and Minami, Yoshiie (2013b) *Iga and Koga Ninja Skills* (Stroud, The History Press).

30. Yamada, Yūji (2014). *Ninja no Kyōkasho: Shin Mansenshūkai*. (Tokyo, Kasamashoin), p. 29
31. Yamada, Yūji (2014). *Ninja no Kyōkasho: Shin Mansenshūkai* (Tokyo, Kasamashoin), p. 18.
32. For a thorough study of Marishiten see Hall, David A. (2013). *The Buddhist Goddess Marishiten* (Leiden, Brill).
33. Yamada, Yūji (2014). *Ninja no Kyōkasho: Shin Mansenshūkai*. (Tokyo, Kasamashoin)., pp. 25-26.
34. Hall, David A. (2013). *The Buddhist Goddess Marishiten* (Leiden, Brill), pp. 285-287.
35. Yamada, Yūji (2014). *Ninja no Kyōkasho: Shin Mansenshūkai*. (Tokyo, Kasamashoin)., p. 26.

Chapter 9
1. Yoshimaru, Katsuya; Yamada, Yūji; Onishi, Yasumitsu *et. al.* (2014). *Ninja bungei kenkyū dokuhon* (Tokyo, Kasamashoin), pp. 50-51.
2. Yoshimaru, Katsuya; Yamada, Yūji; Onishi, Yasumitsu *et. al.* (2014). *Ninja bungei kenkyū dokuhon* (Tokyo, Kasamashoin), pp. 59-60.
3. Halford, Aubrey S. and Halford Giovanna M. (1956). *The Kabuki Handbook* (Rutland, Tuttle), p. 212.
4. Halford, Aubrey S. and Halford Giovanna M. (1956). *The Kabuki Handbook* (Rutland, Tuttle), pp. 216-217.
5. Tanehiko, Ryūtei (1927). *Nise Monogatari Inaka Genji* in the series *Nihon Meichū Zenshū, Vols 20 and 21* (Tokyo).
6. Yoshida Y. (Ed.) (1979). *Taikō ki* Vols 1, 3 and 4 (Tokyo, Kyōikusha), p. 29.
7. Takenouchi Kakusai & Okada Gyokuzan (1802) *Ehon Taikō ki* (Osaka), Vol. 12. http://archive.wul.waseda.ac.jp/kosho/he13/he13_01833/he13_01833_0072/he13_01833_0072.html (Accessed 10 March 2017).
8. The name can also be read Tatsukawa Bunko and appears this way is some bibliographies.
9. Torrance, Richard (2005). 'Literacy and Literature in Osaka, 1890-1940' *Journal of Japanese Studies* 31, p. 54.
10. Gluck, Carol (1985). *Japan's Modern Myths: Ideology in the Late Meiji Period* (Princeton, Princeton University Press), p. 172.
11. Adachi, Ken'ichi (1967). *Taishū geijutsu no fukuryū* (Tokyo, Rironsha). p. 73.
12. Torrance, Richard (2005). 'Literacy and Literature in Osaka, 1890-1940' *Journal of Japanese Studies* 31, p. 57. There are numerous excellent illustrations from the series in Various Authors (2003). *Ninja to Ninjutsu* (Tokyo, Gakken).
13. Torrance, Richard (2005). 'Literacy and Literature in Osaka, 1890-1940' *Journal of Japanese Studies* 31, p. 54.
14. Adachi, Ken'ichi (1957). *Ninjutsu* (Tokyo, Heibonsha), p. 219.
15. Yoshimaru, Katsuya (2014). 'From Shinobi to Ninja: Observations Focusing in the Tatsukawa Bunko Edition of Sarutobi Sasuke' in *The Truth about Ninja (Briefing Notes for lectures)* Mie University Faculty of Humanities, Law and Economics, p. 4.
16. Torrance, Richard (2005). 'Literacy and Literature in Osaka, 1890-1940' *Journal of Japanese Studies* 31, p. 56.

17. Torrance, Richard (2005). 'Literacy and Literature in Osaka, 1890-1940' *Journal of Japanese Studies* 31, p. 54.
18. Williams, Michael P. (2013). 'Early Taishō Japanese Juvenile Pocket Fiction: Tatsukawa Bunko and its Imitators' *University of Pennsylvania Blogs. 24 April 2013.* https://uniqueatpenn.wordpress.com/2013/04/23/early-taisho-japanese-juvenile-pocket-fiction-tachikawa-bunko-and-its-imitators/ (Accessed 12 February 2016).
19. Sekka, Sanjin (1914). *Sarutobi Sasuke* (Osaka, Tachikawa Bunmeidō), p. 1.
20. Sekka, Sanjin (1914). *Sarutobi Sasuke* (Osaka, Tachikawa Bunmeidō), p. 3.
21. Torrance, Richard (2005). 'Literacy and Literature in Osaka, 1890-1940' *Journal of Japanese Studies* 31, p. 56, referring to Sekka, Sanjin (1914). *Sarutobi Sasuke* (Osaka, Tachikawa Bunmeidō), p. 142.
22. Torrance, Richard (2005). 'Literacy and Literature in Osaka, 1890-1940' *Journal of Japanese Studies* 31, p. 55.
23. Sekka, Sanjin (1914). *Sarutobi Sasuke* (Osaka, Tachikawa Bunmeidō), p. 12. Sesshū is Settsu Province, now part of modern Hyōgo Prefecture
24. Thornton, S.A (2007). *The Japanese Period Film: A Critical Analysis* (Jefferson NC, McFarland), p. 92.
25. It may be viewed at https://www.youtube.com/watch?v=Xt9GRKpeDtU (Accessed 10 March 2016).
26. http://movie.walkerplus.com/mv6820/ (Accessed 8 February 2016).

Chapter 10

1. Itō Gingetsu (1909). *Ninjutsu to Yōjutsu translated with notes and additional material by Eric Shahan* (Amazon, 2014), p. 19.
2. Itō Gingetsu (1909). *Ninjutsu to Yōjutsu translated with notes and additional material by Eric Shahan* (Amazon, 2014), pp. 19-20.
3. Itō Gingetsu (1909). *Ninjutsu to Yōjutsu translated with notes and additional material by Eric Shahan* (Amazon, 2014), p. 23.
4. Itō Gingetsu (1909). *Ninjutsu to Yōjutsu translated with notes and additional material by Eric Shahan* (Amazon, 2014), pp. 39-40.
5. Itō Gingetsu (1909). *Ninjutsu to Yōjutsu translated with notes and additional material by Eric Shahan*n (Amazon, 2014), pp. 49-50.
6. Itō Gingetsu (1909). *Ninjutsu to Yōjutsu translated with notes and additional material by Eric Shahan* (Amazon, 2014), p. 57.
7. Itō Gingetsu (1917). *Ninjutsu no gokui translated with notes and additional material by Eric Shahan*ahan (Amazon, 2014), pp. 20-21.
8. Itō Gingetsu (1917). *Ninjutsu no gokui translated with notes and additional material by Eric Shahan*ahan (Amazon, 2014), pp. 20-21.
9. Yamada, Yūji (2016). *Ninja no Rekishi.* (Tokyo, Kadokawa Shoten), pp. 237-238.
10. Itō Gingetsu (1917). *Ninjutsu no gokui translated with notes and additional material by Eric Shahan*ahan (Amazon, 2014), pp. 8-9
11. Itō Gingetsu (1917). *Ninjutsu no gokui translated with notes and additional material by Eric Shahan*ahan (Amazon, 2014), p. 27.
12. Itō Gingetsu (1917). *Ninjutsu no gokui translated with notes and additional material by Eric Shahan*ahan (Amazon, 2014), p. 49.
13. Itō Gingetsu (1917). *Ninjutsu no gokui translated with notes and additional material by Eric Shahan*ahan (Amazon, 2014), p. 72. & pp. 74 & 96.

14. Itō Gingetsu (1937). *Gendaijin no Ninjutsu translated with notes and additional material by Eric Shahanc* (Amazon, 2014).
15. Itō Gingetsu (1937). *Gendaijin no Ninjutsu translated with notes and additional material by Eric Shahanc* (Amazon, 2014), p. 27.
16. Itō Gingetsu (1937). *Gendaijin no Ninjutsu translated with notes and additional material by Eric Shahanc* (Amazon, 2014), pp. 62-63.
17. Itō Gingetsu (1937). *Gendaijin no Ninjutsu translated with notes and additional material by Eric Shahanc* (Amazon, 2014), pp. 66-70 & 121-123.
18. Itō Gingetsu (1937). *Gendaijin no Ninjutsu translated with notes and additional material by Eric Shahanc* (Amazon, 2014), pp. 196-197.
19. Yamada, Yūji (2016). *Ninja no Rekishi.* (Tokyo, Kadokawa Shoten), pp. 237-238.
20. Hevener, Phillip T. (2008). *Fujita Seiko: The Last Koga Ninja* (Xlibris).
21. Fujita Seikō (1936). *Ninjutsu Hiroku* (Tokyo, Chiyoda Shoin), Introduction preceding p. 1.
22. The picture appears as the frontispiece in his own book: Fujita Seikō (1936) *Ninjutsu Hiroku* (Tokyo, Chiyoda Shoin).
23. Anonymous (1964). 'Japan - a good cocktail' *Newsweek* 3 August 1964, p. 31.
24. The clip is entitled 'Demonstration by Fujita Seikō the Last Ninja specially arranged for Bernard Leach' (Marty Gross Film Productions: The Mingei Film Archive Project, Toronto).
25. Leach, Bernard (1978). *Beyond East and West: memoirs, portraits and essays.* (London, Faber), pp. 178-179.
26. It was made into a movie in 1955. http://movie.walkerplus.com/mv24169/ (Accessed 14 May 2017).
27. Suzuki, Shinichi (1949). (translated by Kyoko Seiden) (1996) *Young Children's Talent Education & Its Method* (Van Nuys CA, Alfred Publishing Ltd) pp. 36-37. Note that the translator uses the word ninja.
28. Fujita Seikō (1936). *Ninjutsu Hiroku* (Tokyo, Chiyoda Shoin), p. 149.
29. Fujita, Seikō (1942). *Ninjutsu kara supaisen e* (Tokyo, Tōkuisha); Roley, Don (trans. and ed.) (2015) *Secrets of Koga-ryu ninjutsu: Fujita Seiko's Works on the Art of the Shadow Warrior* (Colorado Springs, Freedom to Excel).
30. Roley, Don (trans. and ed.) (2015). *Secrets of Koga-ryu ninjutsu: Fujita Seiko's Works on the Art of the Shadow Warrior* (Colorado Springs, Freedom to Excel) p 240.
31. Hevener, Phillip T. (2008). *Fujita Seiko: The Last Koga Ninja* (E-book, Xlibris), Location 567.
32. Roley, Don (trans. and ed.) (2015). *Secrets of Koga-ryu ninjutsu: Fujita Seiko's Works on the Art of the Shadow Warrior* (Colorado Springs, Freedom to Excel) p 249.
33. Svinth, Joseph (2002). 'Documentation Regarding the Budo Ban in Japan, 1945-1950' *(Journal of Combative Sports).* http://ejmas.com/jcs/jcsframe.htm (Accessed 20 February 2016).
34. Bennett, Alexander (2015). *Kendo: Culture of the Sword* (Oakland, University of California Press), p. 161.

Chapter 11
1. Hevener, Phillip T. (2008). *Fujita Seiko: The Last Koga Ninja* (E-book, Xlibris), Location 791.
2. Okuse, Heishichirō (1956). *Ninjutsu* (Osaka, Kinki Nippon Tetsudō).
3. Adachi, Ken'ichi (1957). *Ninjutsu* (Tokyo, Heibonsha), p 123.

4. Adachi, Ken'ichi; Ozaki, Hotsuki & Yamada, Munemutsu (1964). *Ninpō: Gendaijin wa naze ninja ni akogareruka* (Tokyo, Sanichi Shinsho).
5. Okuse, Heishichirō (1959). *Ninjutsu Hiden* (Osaka, Bombonsha).
6. Okuse, Heishichirō (1959). *Ninjutsu Hiden* (Osaka, Bombonsha), pp. 115-119.
7. Okuse, Heishichirō (2011). Ninjutsu, sono rekishi to ninja (Tokyo, Jinbutsu), p 158.
8. Adachi, Ken'ichi et.al. (1964). *Ninpō: Gendaijin wa naze ninja ni akogareruka* (Tokyo, Sanichi Shinsho), p. 197.
9. Yamaguchi, M. (1963). *Ninja no Seikatsu.* (Tokyo, Yūzankaku).
10. For a fascinating collection of illustrations from books like these see Various Authors (2003). *Ninja to Ninjutsu* (Tokyo, Gakken).
11. Clements, Jonathan (2016). *A Brief History of the Martial Arts: East Asian Fighting Styles, from Kung Fu to Ninjutsu* (London, Robinson), p. 168.
12. Clements, Jonathan (2016). *A Brief History of the Martial Arts: East Asian Fighting Styles, from Kung Fu to Ninjutsu* (London, Robinson), pp.166-168.
13. Clements, Jonathan (2016). *A Brief History of the Martial Arts: East Asian Fighting Styles, from Kung Fu to Ninjutsu* (London, Robinson). pp. 206-207.
14. This is Tomita Tsuneo's 1948 novel entitled *The 'Ninja' Sarutobi Sasuke* 忍者猿飛佐助. I have not been able to locate a copy, so the reading of the first two characters is unknown.
15. Shibata, Renzaburō (1955). *Sarutobi Sasuke* (Osaka, Tachikawa Bunmeidō), p. 12.
16. http://vintageninja.net/?p=6953 (Accessed 9 March 2017)
17. A recent reprint of the first volume is available as Shirato, Sanpei (2009). *Ninja bugeichō:Kagemaru-den 1* (Tokyo, Kogaku-kan Creative).
18. Yamada, Fūtarō (2014a). *Kōka Ninpōchō* (Tokyo, Kōdansha) and Yamada, Fūtarō (2006) *The Kouga Ninja Scrolls* (New York, Random House Del Ray paperbacks).
19. Yamada, Fūtarō (2014a). *Kōka Ninpōchō* (Tokyo, Kōdansha) and Yamada, Fūtarō (2006) *The Kouga Ninja Scrolls* (New York, Random House Del Ray paperbacks), p. 3.
20. The editions of the novel I have examined are the 1973 and 2003 reprints: Murayama, Tomoyoshi (1973). *Shinobi no mono* (Tokyo, Kadokawa Shoten). Murayama, Tomoyoshi (2003). *Shinobi no mono 1: Jōmaki* (Tokyo, Iwanami Shoten). The latter includes an appendix about the author and his work.
21. Adachi, Ken'ichi (1967). *Taishū geijutsu no fukuryū* (Tokyo, Rironsha), p.65.
22. Murayama, Tomoyoshi (2003). *Shinobi no mono 1: Jōmaki* (Tokyo, Iwanami Shoten), pp. 483-484.
23. Reproduced in http://vintageninja.net/?cat=12 (Accessed 16 January 2017)

Chapter 12

1. This section is based on my own interpretations of the movies themselves, most of which are readily available as DVDs or through the internet. Here I acknowledge the guidance provided by the excellent website *Vintage Ninja*. I have also made good use of the study of popular culture Adachi, Ken'ichi (1967). *Taishū geijutsu no fukuryū* (Tokyo, Rironsha), which suffers only from the limitations that it was being written as the ninja craze began and thus his comments on ninja films are limited. For a fine collection of ninja movie posters see Various Authors (2003). *Ninja to Ninjutsu* (Tokyo, Gakken).
2. The clip is available at https://www.youtube.com/watch?v=eF_X6oZfv70

3. I have been unable to view a copy of this very rare movie and base my conclusions on the extensive collection of stills assembled by the Vintage Ninja website at http://vintageninja.net/?p=3034 and http://vintageninja.net/?p=3078 (Accessed 7 February 2016).
4. Beck, Jerry (2005). *The Animated Movie Guide.* (Chicago Review Press, Chicago). 2005, p. 158.
5. The eight movies are Shinobi no mono 忍びの者（1962); *Zoku Shinobi no mono* 続□忍びの者（1963); *Shin Shinobi no mono* 新□忍びの者（1963); *Shinobi no mono Kirigakure Saizō* 忍びの者 霧隠才蔵（1964 *Shinobi no mono Zoku Kirigakure Saizō* 忍びの者 続□霧隠才蔵（1964）; *Shinobi no mono Iga yashiki* 忍びの者 伊賀屋敷（1965）; *Shinobi no mono Shin Kirigakure Saizō* 忍びの者 新□霧隠才蔵（1966); *Shinsho Shinobi no mono* 新書□忍びの者（1966)
6. Adachi, Ken'ichi (1967). *Taishū geijutsu no fukuryū* (Tokyo, Rironsha), p. 67.
7. Most of the information for this section comes from an interview with Kawakami Jinichi, Honorary Director of the Iga-ryū Ninja museum on 27 February 2014, together with material collected at the museum itself and observations of its collections between 1986 and 2014.
8. Anonymous (1964). 'Japan - a good cocktail' *Newsweek* 3 August 1964, p. 34.
9. The clip is available at https://www.youtube.com/watch?v=F7A-EJ_trFg. (Accessed 21 February 2016).
10. Anonymous (1964). 'Japan - a good cocktail' *Newsweek* 3 August 1964, p. 31.
11. Keith Rainville has provided an excellent overview of *You Only Live Twice* to commemorate its fiftieth anniversary. See http://vintageninja.net/?p=8759. (Accessed 28 June 2017).
12. Fleming, Ian (1963). *Thrilling Cities* (London, Jonathan Cape), pp, 51-71.
13. Fleming, Ian (1964). *You Only Live Twice* (London, Jonathan Cape), p 95.
14. http://vintageninja.net/?p=8759. (Accessed 28 June 2017).
15. https://www.animeigo.com/liner/samurai/shinobi-no-mono/ (Accessed 9 June 2017).
16. Information derived from the optional voice-over commentary added to the film on its DVD boxed set.
17. As proven pictorially in http://vintageninja.net/?p=8759. (Accessed 28 June 2017).
18. Apart from discerning readers of the *Daily Express* of course!
19. Adams, Andy (1966). 'A Leap into the Supernatural' *Black Belt 4, 12* December 1966 pp. 12-21; Adams, Andy (1967a). 'A Curriculum for Assassins' *Black Belt 5, 1* January 1967 pp. 16-29; Adams, Andy (1967b). 'The Last of the Ninja' *Black Belt 5, 2* February 1967 pp. 26-33.
20. Adams, Andrew (1970). *Ninja: the invisible assassins* (Burbank, Ohara Publications), pp. 11-12.
21. Adams, Andy (1967c). '007's newest "gimmick" a whole arsenal of Japanese self-defense arts' *Black Belt* 5, 8 August 1967 pp. 34-39.
22. Clements, Jonathan (2016). *A Brief History of the Martial Arts: East Asian Fighting Styles, from Kung Fu to Ninjutsu* (London, Robinson), pp. 208-209.

Chapter 13
1. Takao, Yoshiki (2017). *Ninja no Matsue: Edo-jō ni tsutometa Iga-mono tachi* (Tokyo, Kadokawa Shoten).

2. Sasama, Yoshihiko (1968). *Buke Senjin Sakuhō Shūsei*. (Tokyo, Yūzankaku), pp. 384-385.
3. Imamura, Yoshio (Ed.) (1966) *Nihon budō zenshū Volume 2* (Tokyo, Jinbutsu Oraisha), p. 94.
4. Reproduced in Watatani, Kiyoshi (1972). *Bugei ryūha hyakusen* (Tokyo, Akita Shoten), pp. 229-232 and in Various Authors (2017b). *Ninja no Dokuhon* (Tokyo, Takarajimasha), p. 23.
5. Various Authors (2017b). *Ninja no Dokuhon* (Tokyo, Takarajimasha), p. 25.
6. Fujita Seikō (1936). *Ninjutsu Hiroku* (Tokyo, Chiyoda Shoin), p. 193.
7. Fujita Seikō (1936). *Ninjutsu Hiroku* (Tokyo, Chiyoda Shoin), p. 194. The picture is reproduced in the plate section of this book.
8. Kawakami, Jinichi (Ed.). (1998) *Iga-ryū kaki hi no maki* (Iga City, Iga-ryū Ninja Museum).
9. Naruse, Kanji (1943). *Shūriken* (Tokyo, Shin Ōshūsha), pp. 200-202.
10. Naruse, Kanji (1943). *Shūriken* (Tokyo, Shin Ōshūsha), pp. 165-166. The anecdote is also related in Shiragami, Ikkūken (2001) *Shūriken no sekai* (Tokyo, Sōjinsha), pp. 59-60.
11. Nakashima, Atsumi (2016). *Ninja o kagaku suru* (Tokyo, Yōsensha), p. 80.
12. Kawakami, Jinichi (Ed.) (1998). *Iga-ryū kaki hi no maki* (Iga City, Iga-ryū Ninja Museum).
13. Nakashima, Atsumi (2016). *Ninja o kagaku suru* (Tokyo, Yōsensha), pp. 136-137.
14. Nakashima, Atsumi (2016). *Ninja o kagaku suru* (Tokyo, Yōsensha), pp. 134-135.
15. Okuse, Heishichirō (1956). *Ninjutsu* (Osaka, Kinki Nippon Tetsudō), p. 21.
16. Adachi, Ken'ichi et. al. (1964). *Ninpō: Gendaijin wa naze ninja ni akogareruka* (Tokyo, Sanichi Shinsho), p.213.
17. http://vintageninja.net/?p=6953 (Accessed 9 March 2017)
18. Adachi, Ken'ichi (1967). *Taishū geijutsu no fukuryū* (Tokyo, Rironsha). p.10.
19. Fujita Seikō (1966). *Zukai Shūriken-jutsu* (Tokyo, Inoue Tosho).
20. Fujita Seikō (1966). *Zukai Shūriken-jutsu* (Tokyo, Inoue Tosho), p. 61.
21. Higuchi, Naofumi (2008). *'Gekkō kamen' o hajimeta otokotachi* (Tokyo, Heibonsha Shinsho), p. 165.
22. Reproduced on http://vintageninja.net/?p=8173 (Accessed 21 February 2016).
23. Jeng, Bennie H. (et al) (2001). 'Severe Penetrating Ocular Injury From Ninja Stars in Two Children' *Ophthalmic Surgery, Lasers and Imaging Retina* 32, 4, pp. 336-337.

Chapter 14
1. For some strange reason the 2017 movie *Shinobi no Kuni* was released in English-speaking countries as *Mumon: Land of Stealth*. Mumon is the hero in the film and 'Land of Stealth' is a curious translation of *Shinobi no Kuni*, thereby abandoning any connection with ninja whatsoever!
2. http://www3.nhk.or.jp/lnews/tsu/3073533901.html?t=1485927394630 (Accessed 2 February 2017).
3. Ikeda, Hiroshi (2003). *Iga Ninja: 49 true stories* (Iga City: Igabito no Omoi Jitsugen Jinkai).
4. Ninjaemon, an ellipsoidal ninja dressed in black, must be distinguished from Ninjamon, a ninja-like character dressed in red who has no connection with either Iga or Kōka and Ninja Emon, a manga figure.

5. Yamada, Yūji (2014). *Ninja no Kyōkasho: Shin Mansenshūkai.* (Tokyo, Kasamashoin), p. 8.
6. From a personal observation made on 31 January 2015. See also Yamada, Yūji (2014). *Ninja no Kyōkasho: Shin Mansenshūkai.* (Tokyo, Kasamashoin), pp. 36-38.
7. The Kawagoe Historical Museum also owns one of the finest private collections of genuinely old samurai armour in the whole of Japan and a visit is highly recommended.
8. Various Authors (2017b). *Ninja no Dokuhon* (Tokyo, Takarajimasha), pp. 32-33.
9. Mosher, Gouverneur (1964). *Kyoto: A Contemplative Guide.* (Rutland VT, Tuttle), pp. 181-190 & 311-318.
10. From a personal observation made on 16 February 2015. Yes of course I enjoyed it.
11. http://ninja-do.com/ (Accessed 21 February 2017).
12. http://www.hizenyumekaidou.info/ (Accessed 21 February 2017).
13. There is a cheerful accompanying book: Shimizu, Torazō (1982). *Togakushi no ninja* (Nagano, Ginga Shobo).
14. From personal observations made on 5 July 2017.
15. From a personal observation made on 25 April 2017.

Chapter 15

1. https://ninja-official.com/ (Accessed 2 June 2017).
2. See, for example, Parry, Richard Lloyd (2016). 'Wanted: a team of ninjas to kill time for tourists' *The Times*, 15 March 2016.
3. Clements, Jonathan (2016). *A Brief History of the Martial Arts: East Asian Fighting Styles, from Kung Fu to Ninjutsu* (London, Robinson), p. 213.
4. Various Authors (2016). *The Ninja - ninjatte nanja!? - kōshiki bukku* (Tokyo, Asahi Shimbun).
5. Nakashima, Atsumi (2016). *Ninja o kagaku suru* (Tokyo, Yōsensha).
6. Nakashima, Atsumi (2015). *Mansenshūkai* (Tokyo, Kokusho); Nakashima, Atsumi (2017) *Ninja no heihō* (Tokyo, Kadokawa Shoten).
7. Various Authors (2017c). *Mie Daigaku Iga Kenkyū Kyoten: Kenkyū - Katsudō Gaiyō - Ninja, Ninjutsu ni kansuru tokubetsugō* (Iga City, Mie University).
8. For a brief summary see Various Authors (2016). *The Ninja - ninjatte nanja!? - kōshiki bukku* (Tokyo, Asahi Shimbun), p. 102-107. His work has recently been published as Komori, Teruhisa (2017). *Ninja 'makenai kokoro' no himitsu.* (Tokyo Seishun Shuppansha).
9. Kawakami, Jinichi (2016). *Ninja no okite* (Tokyo: Kadokawa Shoten), p. 8.
10. From an interview with the author on 27 February 2014.
11. China Internet Information Center, Japanese Edition (China Net) 14 February 2012. In Japanese in Yamada, Yūji (2016) *Ninja no Rekishi.* (Tokyo, Kadokawa Shoten), pp. 257-258. The article has been translated into English in Yamada, Yūji (2014) *The Spirit of Ninja: A Study of the Global Ninja Craze* (Tsu City, Mie University) E-book, Loc. 22.
12. Yamada, Yūji (2014). *The Spirit of Ninja: A Study of the Global Ninja* Craze (Tsu City, Mie University) E-book, Loc. 36.
13. Kawakami, Jinichi (2016). *Ninja no okite* (Tokyo: Kadokawa Shoten), p. 23
14. Kawakami, Jinichi (2016). *Ninja no okite* (Tokyo: Kadokawa Shoten), p. 205.
15. Kawakami, Jinichi (2016). *Ninja no okite* (Tokyo: Kadokawa Shoten), p. 23

16. Kawakami, Jinichi (2016). *Ninja no okite* (Tokyo: Kadokawa Shoten), p. 26-27.
17. Kawakami, Jinichi (2016) *Ninja no okite* (Tokyo: Kadokawa Shoten), p. 201.
18. Kawakami, Jinichi (2016) *Ninja no okite* (Tokyo: Kadokawa Shoten), p. 205.
19. Kawakami, Jinichi (2014).'The Body and Mind of a Ninja' Briefing notes for the lecture *The Truth About Ninja* Japan Foundation, London November 2014, p. 2.
20. Kawakami, Jinichi (2016). *Ninja no okite* (Tokyo: Kadokawa Shoten), p. 205.
21. Hayes, Stephen K (1985) *The Mystic Arts of the Ninja: Hypnotism, Invisibility and Weaponry* (Chicago, Contemporary Books), p. 9.
22. Griffith, Samuel B. (trans.) (1963). *Sun Tzu: The Art of War* (Oxford, Oxford University Press), p. 79.
23. Yamada, Yūji (2016) *Ninja no Rekishi*. (Tokyo, Kadokawa Shoten), p. 13.
24. Hobsbawm, E. & Ranger, T. (Eds.) (1983). *The Invention of Tradition*. (Cambridge University Press), p. 1.
25. Hobsbawm, E. & Ranger, T. (Eds.) (1983). *The Invention of Tradition*. (Cambridge University Press), p. 4.
26. Hobsbawm, E. & Ranger, T. (Eds.) (1983). *The Invention of Tradition*. (Cambridge University Press), pp. 4 & 7.
27. The Naruto phenomenon is discussed in Yamada, Yūji (Ed.) (2014d) *The Ninja Book: the new Mansenshūkai* (Tsu City, Mie University) E-book.

Bibliography

Adachi, Ken'ichi (1957). *Ninjutsu* (Tokyo, Heibonsha).

Adachi, Ken'ichi; Ozaki, Hotsuki & Yamada, Munemutsu (1964). *Ninpō: Gendaijin wa naze ninja ni akogareruka* (Tokyo, Sanichi Shinsho).

Adachi, Ken'ichi (1967). *Taishū geijutsu no fukuryū* (Tokyo, Rironsha).

Adams, Andy (1966). 'A Leap into the Supernatural' *Black Belt 4, 12* December 1966 pp. 12-21.

Adams, Andy (1967a). 'A Curriculum for Assassins' *Black Belt 5, 1* January 1967 pp. 16-29.

Adams, Andy (1967b). 'The Last of the Ninja' *Black Belt 5, 2* February 1967 pp. 26-33.

Adams, Andy (1967c). '007's newest "gimmick" a whole arsenal of Japanese self-defense arts' *Black Belt 5,* 8 August 1967 pp. 34-39.

Adams, Andrew (1970). *Ninja: the invisible assassins* (Burbank, Ohara Publications).

Adolphson, Mikael & Commons, Anne (eds.) (2015). *Loveable Losers: The Heike in Action and Memory* (Honolulu, Univ of Hawaii Press).

Akita, Hiroki (1990). *Oda Nobunaga to Azuchijō* (Osaka, Sōgensha).

Anonymous (1891). *Tourist's Guide and Interpreter* (Tokyo, Seishi Bunsha).

Anonymous (1954). *Nagel's Travel Guide to Japan* (Geneva, Nagel Publishers)

Anonymous (1964). 'Japan - a good cocktail' *Newsweek* 3 August 1964, pp. 31-34.

Anonymous (1966). 'Editorial' *Black Belt,* 4, 12 December 1966, p. 11.

Araki, Eishi. (1987). *Higo Kunishū ikki* (Kumamoto, Kumamoto Shuppan Bunka Kaikan).

Asawaka, K. (1929). *The Documents of Iriki: Illustrative of the Development of the Feudal Institutions of Japan.* (Yale University Press).

Aston W.G. (1896) 'Nihongi: Chronicles of Japan from the Earliest times to A.D. 697 Volume II' *Transactions and Proceedings of the Japan Society of London* Supplement I.

Aston, W.G. (1972). *Nihongi: Chronicles of Japan from the Earliest Times to A.D. 697* (Volume I and II) (Reprint). (Rutland,Vermont, Tuttle).

217

Beck, Jerry (2005). *The Animated Movie Guide*. (Chicago Review Press, Chicago). 2005.

Benesch, Oleg (2014). *Inventing the Way of the Samurai: Nationalism, Internationalism, and Bushidō in Modern Japan* (Oxford, Oxford University Press).

Bennett, Alexander (2015). *Kendo: Culture of the Sword* (Oakland, University of California Press).

Berry, Elizabeth Mary (1982). *Hideyoshi* (Cambridge MA, Harvard University Press).

Birt, Michael Patrick. (1983). *Warring States: A study of the Go-Hōjō daimyo and domain 1491-1590* (Ph.D Thesis, Princeton University).

Blair, Emma Helen & Robertson, James Alexander (Eds.) (1903-9). *The Philippine Islands, 1493-1803: (Volume 1 Part 24.)* (Cleveland, Arthur H. Clark Co.).

Clements, Jonathan & McCarthy, Helen (2006). *The anime encyclopaedia: a guide to Japanese animation since 1917* (Berkeley, Stone Bridge Press).

Clements, Jonathan (2013). *Anime: A History* (London, Palgrave Macmillan).

Clements, Jonathan (2016). *A Brief History of the Martial Arts: East Asian Fighting Styles, from Kung Fu to Ninjutsu* (London, Robinson).

Cooper, Michael (1974). *Rodrigues the Interpreter: An Early Jesuit in Japan and China* (New York, Weatherhill).

Cummins, Antony and Minami, Yoshiie (2011). *True Path of the Ninja* (Rutland VT, Tuttle).

Cummins, Antony and Minami, Yoshiie (2012). *The Secret Traditions of the Shinobi* (Berkeley, Blue Snake Books).

Cummins, Antony and Minami, Yoshiie (2013a). *The Book of Ninja* (London, Watkins Press).

Cummins, Antony and Minami, Yoshiie (2013b). *Iga and Koga Ninja Skills* (Stroud, The History Press).

Diosy, Arthur (1911). 'Yoshitsune, the boy hero of Japan ' *Transactions and Proceedings of the Japan Society of London 10*, pp. 50-77.

Doi, Tadao (Ed.) (1960). *Nippo jisho: Vocabvlario da lingo de Iapam*. (Tokyo, Iwanami Shōten).

Elison, George. (1981). 'The Cross and the Sword: Patterns of Momoyama History' in George Elison and Bardwell L. Smith (Eds.) *Warlords, Artists and Commoners: Japan in the Sixteenth Century* (University of Hawaii Press, Honolulu).

Elisonas, Jurgis (1991). 'Christianity and the daimyo' In Hall, J. W. & McLain, J. L. (Eds.) *The Cambridge History of Japan. Vol. 4 Early modern Japan*. (Cambridge University Press), pp. 301-372).

Elisonas, J.S.A. & Lamers, J.P. (Trans. and Ed.) (2011). *The Chronicle of Lord Nobunaga by Ōta Gyūichi*. (Leiden, Brill).

Ferejohn, J.A. & Rosenbluth, F.M. (Eds.) (2010). *War and State Building in Medieval Japan*. (Stanford University Press).

Friday, Karl (2004). *Samurai, warfare and the state in early medieval Japan* (New York, Routledge).

Fujiki, Hisashi (2005). *Katanagari* (Tokyo, Iwanami Shoten).

Fujiki, Hisashi (2005). *Zōhyōtachi no senjō: chūsei no yōhei to dorei kari* (Tokyo, Asahi Shinbun).

Fujita, Kazutoshi (2012). *'Kōka ninja' no jitsuzō* (Tokyo, Yoshikawa Kōbunkan).

Fujita, Seikō (1936). *Ninjutsu Hiroku* (Tokyo, Chiyoda Shoin).

Fujita, Seikō (1942). *Ninjutsu kara supaisen e* (Tokyo, Tōkuisha).

Fujita, Seikō (1966). *Zukai Shūrikenjutsu* (Tokyo, Inoue Tosho).

Fukai, Masaumi (1992). *Edojō oniwaban : Tokugawa Shōgun no mimi to me* (Tokyo, Chūō Kōronsha).

Gluck, Carol (1985). *Japan's Modern Myths: Ideology in the Late Meiji Period* (Princeton, Princeton University Press).

Griffith, Samuel B. (trans.) (1963). *Sun Tzu: The Art of War* (Oxford, Oxford University Press).

Hagiwara, T. (Ed.) (1966). *Hōjō Godai ki in Sengoku Shiryō Sōsho Vol. 21.* (Tokyo: Jinbutsu Oraisha).

Halford, Aubrey S. and Halford Giovanna M. (1956). *The Kabuki Handbook* (Rutland, Tuttle).

Hall, David A. (2013). *The Buddhist Goddess Marishiten* (Leiden, Brill).

Hall, John Whitney (Ed.) (1981). *Japan Before Tokugawa: Political Consolidation and Economic Growth, 1500 to 1650* (Princeton University Press).

Hanawa, Hoki(no)ichi (1960). *Gunsho Ruijū Volume 27* (kassen bu) (Tokyo, Zoku Gunsho Ruijū Kanseikai).

Harada, T. & Watanabe, M. (1972). *Shiga-ken no rekishi.* (Tokyo: Yamakawa Shuppansha).

Hatai, Hiromu (1975). *Shugo ryōgoku taisei no kenkyū : Rokkaku Shi ryōgoku ni miru Kinai kinkokuteki hatten no tokushitsu.* (Tokyo, Yoshikawa Kōbunkan).

Hayashi, Senkichi (1954). *Shimabara Hantō-shi Volume II.* (Nagasaki, Minami Takaki-gun Board of Education).

Hayes, Stephen K (1985). *The Mystic Arts of the Ninja: Hypnotism, Invisibility and Weaponry* (Chicago, Contemporary Books),

Hevener, Phillip T. (2008). *Fujita Seiko: The Last Koga Ninja* (Xlibris).

Higuchi, Naofumi (2008). *'Gekkō kamen' o hajimeta otokotachi* (Tokyo, Heibonsha Shinsho).

Hikone City Board of Education (no date). *'Watashi no machi no Sengoku': Dodo uji to Dodo yakata* (information sheet) (Hikone City).

Hobsbawm, E. & Ranger, T. (Eds.) (1983). *The Invention of Tradition.* (Cambridge University Press).

Hori, Ichiro. (1966). 'Mountains and their importance for the idea of the other world in Japanese folk religion' *History of Religions 6*, pp. 1-23.

Iga City (2011). *Iga-shi shi Volume 1* (Iga City).

Iguchi, Asao (1995). *Maeda Toshiie* (Tokyo, Seibido Shuppan).

Ikeda, Hiroshi (2003). *Iga Ninja: 49 true stories* (Iga City: Igabito no Omoi Jitsugen Jinkai).

Imamura, Yoshio (Ed.) (1966). *Nihon budō zenshū Volume 4* (Tokyo, Jinbutsu Oraisha).

Imamura, Y. (Ed.) (2005). *Ōu Eikei Gunki.* (Tokyo, Jinbutsu Oraisha).

Itō Gingetsu (1909). *Ninjutsu to Yōjutsu translated with notes and additional material by Eric Shahan* (Amazon, 2014).

Itō Gingetsu (1917). *Ninjutsu no gokui translated with notes and additional material by Eric Shahan* (Amazon, 2014).

Itō Gingetsu (1937). *Gendaijin no Ninjutsu translated with notes and additional material by Eric Shahan* (Amazon, 2014).

Iwao, Seiichi (1934). 'Matsukura Shigemasa no Ruzonto ensei keikaku' *Shigaku Zasshi* 45, 9, pp. 81-109.

Jansen, Marius B. (1995). *Warrior Rule in Japan* (New York, Cambridge University Press).

Jeng, Bennie H. (et al) (2001). 'Severe Penetrating Ocular Injury From Ninja Stars in Two Children' *Ophthalmic Surgery, Lasers and Imaging Retina* 32, 4, pp. 336-337.

Kawakami, Jinichi (2016). *Ninja on okite* (Tokyo: Kadokawa Shoten).

Kawakami, Jinichi (Ed.) (1998). *Iga-ryū kaki hi no maki* (Iga City, Iga-ryū Ninja Museum).

Kido, Masayuki (2008). 'O[da]Toyo[tomi]ki no Kōka: Kōka no yakiuchi wa nakatta' *Kiyō* (Shiga Bunkazai) 21, pp. 34-46.

Kikuoka, Jōgen (2006). *Iga Kyūkō Iran ki* (Iga City, Iga Kobunken Kankōkai).

Kōka City Board of Education (2011). *Wada jōkan gun.* (Kōka no jōkaku 1) (Kōka City).

Komori, Teruhisa (2017). *Ninja 'makenai kokoro' no himitsu.* (Tokyo Seishun Shuppansha).

Konstam, Angus (2011). *Pirate: The Golden Age.* (Oxford, Osprey Publishing).

Kōryō Town Historical Records Committee (2000). *Kōryō-chō shi Volume 1.* (Kōryō Town, Nara Prefecture).

Kōyasan Shihensan Shohen (1936). *Kōyasan Bunsho Volume 9* (Kōyasan Bunsho Kankō).

Kumamoto City. (2000). *Kumamoto-shi shi kankei shiryōshū. Volume 4: Higo koki shūran.* (Kumamoto City).

Kurushima, Noriko.(2011). *Ikki no sekai to hō* (Tokyo, Yamakawa).

Kuwata, Tadachika (Ed.) (1943). *Taikō ki.* (Tokyo, Iwanami Shoten).

Kuwata, Tadachika. (1975). *Toyotomi Hideyoshi Kenkyū.* (Tokyo, Kadokawa Shoten).

Kuwata, Tadachika & Utagawa, T. (Eds.) (1976). *Kaisei Mikawa Gofudo ki*. (Tokyo, Akita Shoten).

Lamers, J. P. (2000). *Japonius Tyrannus: The Japanese Warlord Oda Nobunaga Reconsidered*. (Leiden, Hotei Publishing).

Leach, Bernard (1978). *Beyond East and West: memoirs, portraits and essays*. (London, Faber).

Lemagnen, Guillaume (2015). *Demystifying Ninjutsu, a Necessary Task* (Amazon).

Manser, Martin H. (2011). *Concise English Chinese Chinese-English Dictionary 4th Edition* (Beijing, Xianggang, Shang wu yin shu guan).

Marcure, Kenneth (1985). 'The Danka System' *Monumenta Nipponica* 40, 1, pp. 39-67.

Matsubara Kazuyoshi (2005). *'Mogami Yoshimitsu Monogatari no kaisetsu to honkoku Part 2' Bulletin of Naruto Kyoiku University Human and Social Sciences* Vol. 20, pp. 1-10.

McCullough, Helen Craig (1959). *The Taiheiki: A Chronicle of Medieval Japan*. (New York, Columbia University Press).

Momochi, Orinosuke (1897). *Kōsei Iran ki* (Iga-Ueno).

Morimoto, Masahiro (2014). *Kyōkai arasoi to Sengoku chōhōsen*. (Tokyo, Yōsensha).

Morinaga, Maki Isaka (2005). *Secrecy in Japanese Arts: "Secret Transmission" as a Mode of Knowledge* (New York, Palgrave Macmillan).

Mosher, Gouverneur (1964). *Kyoto: A Contemplative Guide*. (Rutland VT, Tuttle).

Murayama, Tomoyoshi (1973). *Shinobi no mono* (Tokyo, Kadokawa Shoten).

Murayama, Tomoyoshi (2003). *Shinobi no mono 1: Jōmaki* (Tokyo, Iwanami Shoten).

Nakashima, Atsumi (2015). *Mansenshūkai* (Tokyo, Kokusho).

Nakashima, Atsumi (2016). *Ninja o kagaku suru* (Tokyo, Yōsensha).

Nakashima, Atsumi (2017). *Ninja no heihō* (Tokyo, Kadokawa Shoten).

Nara Joshi Daigaku (1991). *Nara Joshi Daigaku kyōiku kenkyū nai tokubetsu keihi (Nara bunka ni kansuru sōgōteki kenkyū) hōkokusho*. (Nara, Nara Women's University).

Naruse, Kanji (1943). *Shūriken* (Tokyo, Shin Ōshūsha).

Nelson, Andrew N. (1997). *The New Nelson Japanese-English Character Dictionary; based on the Classic Edition by Andrew N. Nelson, completely revised by John H. Haig*. (Rutland, Tuttle).

Nitobe, Inazu (1905). *Bushido: The Soul of Japan: Tenth Revised Edition*. (New York, Putnam's).

Norman, E. Herbert (1940). *Japan's Emergence as a Modern State: Political and Economic Problems of the Meiji Period* (New York, Institute of Pacific Relations).

Okada, Akio (1960). *Amakusa Tokisada*. (Tokyo, Yoshikawa Kōbunkan).

Okuno, Takahiro (1960). *Ashikaga Yoshiaki* (Tokyo, Yoshikawa Kōbunkan).

Okuno Takahiro (1988). *Oda Nobunaga monjo no kenkyū* (Tokyo, Yoshikawa Kōbunkan).

Okuse, Heishichirō (1956). *Ninjutsu* (Osaka, Kinki Nippon Tetsudō).

Okuse, Heishichirō (1959). *Ninjutsu Hiden* (Osaka, Bombonsha).

Okuse, Heishichirō (1964). *Ninpō* (Tokyo, Jinbutsu Ōraisha).

Okuse, Heishichirō (2011). *Ninjutsu, sono rekishi to ninja* (Tokyo, Jinbutsu).

Owada, Tetsuo (1973). *Ōmi Azai Shi* (Tokyo, Shin Jinbutsu Oraisha).

Owada, Tetsuo (2006). *Hideyoshi no tenka tōitsu sensō* (Tokyo, Yoshikawa Kobunkan).

Ōyama, Ryūshū (2003). *Hideyoshi to Higo Kunishū ikki* (Fukuoka, Kaichosha).

Oxenboell, Morten (2005). 'Images of "Akutō"' *Monumenta Nipponica*, 60, 2, pp 235-262.

Oxenboell, Morten (2006). 'Mineaiki and Discourses on Social Unrest in Medieval Japan' *Japan Forum* 18, 1, pp. 1-21.

Parry, Richard Lloyd (2016). 'Wanted: a team of ninjas to kill time for tourists' The Times, 15 March 2016.

Rabinovitch, Judith (1986). *Shōmonki, the story of Masakado's rebellion* (Tokyo, Monumenta Nipponica Monographs)

Roberts, Luke (1997). 'A Petition for a Popularly Chosen Council of Government in Tosa in 1787'. *Harvard Journal of Asiatic Studies* 57, 2 pp. 575-596.

Roley, Don (trans. and ed.) (2015). *Secrets of Koga-ryu ninjutsu: Fujita Seiko's Works on the Art of the Shadow Warrior* (Colorado Springs, Freedom to Excel).

Sasama, Yoshihiko (1968). *Buke Senjin Sakuhō Shūsei.* (Tokyo, Yūzankaku).

Schuessler, Axel (2007). *ABC Etymological Dictionary of Old Chinese* (Honolulu, Hawaii University Press, 2007).

Schwade, Arcadio (1964). 'Matsukura Shigemasa no Ruzonto ensei keikaku' *Kirishitan Kenkyū* 9, pp. 337-350.

Sekka, Sanjin (1914). *Sarutobi Sasuke* (Osaka, Tachikawa Bunmeidō).

Shibata, Renzaburō (1955). *Sarutobi Sasuke* (Osaka, Tachikawa Bunko).

Shimizu, Noboru (2008). *Sengoku ninja retsuden* (Tokyo, Kawade Shobo Shinsha).

Shimizu, Noboru (2009). *Sengoku ninja wa rekishi o dō ugokashita no ka?* (Tokyo, Best Shinsho).

Shimizu, Noboru (2009). *Edo no onmitsu: oniwaban* (Tokyo, Kawade Shobo Shinsha).

Shimizu, Torazō (1982). *Togakushi no ninja* (Nagano, Ginga Shobo).

Shinoda, Minoru (1960). *The Founding of the Kamakura Shogunate 1180-1185* (New York, Columbia University).

Shinya, Kazuyuki (2015). *Ōmi Rokkaku shi.* (Tokyo, Ebisu-Kosho).

Shiragami, Ikkūken (2001). *Shūriken no sekai* (Tokyo, Sōjinsha).

Shirato, Sanpei (2009). *Ninja bugeichō:Kagemaru-den 1* (Tokyo, Kogaku-kan Creative).

Souyri, Pierre. (2001). *The World Turned Upside Down: Medieval Japanese Society* (Columbia University Press).

Souyri, P. (2010). 'Autonomy and War in the Sixteenth Century Iga Region and the Birth of the Ninja Phenomena', in Ferejohn, J. A. & Rosenbluth, F. M. (Eds) *War and State Building in Medieval Japan.* (Stanford University Press), pp. 110-123.

Sugiyama, Hiroshi (1974). *Sengoku Daimyo*. (Tokyo, Chūō Kōronsha).

Suzuki, Shinichi (1949). (translated by Kyoko Seiden) (1996) *Young Children's Talent Education & Its Method* (Van Nuys CA, Alfred Publishing Ltd).

Svinth, Joseph (2002). 'Documentation Reagarding the Budo Ban in Japan, 1945-1950' *(Journal of Combative Sports)*. http://ejmas.com/jcs/jcsframe.htm (Accessed 20 February 2016).

Takao, Yoshiki (2017). *Ninja no Matsue: Edo-jō ni tsutometa Iga-mono tachi* (Tokyo, Kadokawa).

Takenouchi Kakusai & Okada Gyokuzan (1802). *Ehon Taikō ki* (Osaka).

Takeuchi, Rizō (Ed.) (1978). *Tamon'In Nikki (Zōho Zoku Shiryō taisei Volume 38)* (Rinsen Shoten, Kyoto).

Tamada, Gyokushūsai & Yamada, Tadao (1910). *Sarutobi Sasuke* (Osaka, Matsumoto).

Tanehiko, Ryūtei (1927). *Nise Monogatari Inaka Genji in the series Nihon Meichū Zenshū, Vols 20 and 21* (Tokyo).

Tokyo Daigaku Shiryō Hensansho (1975). *Dai Nihon Shiryō Part 11, Vol 15*. Tokyo: Tokyo University. (Cited as DNS).

Thornton, S.A (2007). *The Japanese Period Film: A Critical Analysis* (Jefferson NC, McFarland).

Toda, Toshio (1988). *Amakusa-Shimabara no ran: Hosokawa han shiryō ni yoru*. (Tokyo: Shin Jinbutsu Ōraisha).

Torrance, Richard (2005). 'Literacy and Literature in Osaka, 1890-1940' *Journal of Japanese Studies* 31, pp. 27-60.

Tsuruta, Sōzō (1981). 'Tenshō Amakusa Kassen no kōsatsu' *Kumamoto Shigaku* 55, pp. 47-58.

Turnbull, Stephen (1991). *Ninja: The True Story of Japan's Secret Warrior Cult*, (London, Firebird Books).

Turnbull, Stephen (1992). 'Pictures of Invisible Men' *Andon* 41, pp. 13-18.

Turnbull, Stephen (2003). *Ninja: AD 1460-1650* (Oxford, Osprey Publishing Ltd).

Turnbull, Stephen (2007). *Ninja: Die Wahre Geschichte Der Geheimnisvollen Japanischen Schattenkrieger* (Zürich, Motorbuch Verlag).

Turnbull, Stephen (2008). *Real Ninja* (New York, Enchanted Lion Books).

Turnbull, Stephen (2013). 'The ghosts of Amakusa: localised opposition to centralised control in Higo Province, 1589-90.' *Japan Forum* 25 (2), pp. 191-211.

Turnbull, Stephen. (2014). 'The Ninja: An Invented Tradition?' *Journal of Global Initiatives: Policy, Pedagogy, Perspective: Interdisciplinary Reflections on Japan Vol. 9 No. 1*, Article 3, pp. 9-26.

Turnbull, Stephen (2016). 'Wars and Rumours of Wars: Japanese Plans to invade the Philippines, 1593-1637' *Naval War College Review Vol. 69, no, 4*, pp. 107-120.

Turnbull, Stephen (in preparation). *Ninja: The (Unofficial) Secret Manual* (London, Thames and Hudson).

Ueda, Mitsue (2015). *Ninja no Dietto* (Tokyo, Sideranch,ink).

Urban, William (2006). *Medieval Mercenaries: The Business of War*. (London, Greenhill Books).

Van Bremen, Jan and Martinez, D.P. (eds.) (1995). *Ceremony and Ritual in Japan: Religious practices in an industrialised society*. (London, Routledge).

Various Authors (2002). *Iga.Kōka shinobi no subete* Bessatsu Rekishi Dokuhon 619. (Tokyo, Shinjinbutsu).

Various Authors (2003). *Ninja to Ninjutsu* (Tokyo, Gakken).

Various Authors (2016). *The Ninja - ninjatte nanja!? - kōshiki bukku* (Tokyo, Asahi Shimbun).

Various Authors (2017a). *Ninja to yōkai* (Kwai no. 50) (Tokyo, Kadokawa Shoten).

Various Authors (2017b). *Ninja no Dokuhon* (Tokyo, Takarajimasha).

Various Authors (2017c). *Mie Daigaku Iga Kenkyū Kyoten: Kenkyū - Katsudō Gaiyō - Ninja, Ninjutsu ni kansuru tokubetsugō* (Iga City, Mie University).

Varley, H. Paul (1994). *Warriors of Japan as portrayed in the war tales* (Honolulu, University of Hawaii Press)

Vlastos, Stephen (Ed.) (1998). *Mirror of Modernity: Invented Traditions of Modern Japan*. (Berkeley, University of California Press).

Wada, Ryo (2008). *Ninja no Kuni* (Tokyo, Shinchosa).

Watatani, Kiyoshi (1972). *Bugei ryūha hyakusen* (Tokyo, Akita Shoten).

Williams, Michael P. (2013). 'Early Taishō Japanese Juvenile Pocket Fiction: Tatsukawa Bunko and its Imitators' *University of Pennsylvania Blogs. 24 April 2013.* https://uniqueatpenn.wordpress.com/2013/04/23/early-taisho-japanese-juvenile-pocket-fiction-tachikawa-bunko-and-its-imitators/ (Accessed 12 February 2016).

Yamada, Fūtarō (2006). *The Kouga Ninja Scrolls* (New York, Random House Del Ray paperbacks).

Yamada, Fūtarō (2012). *Kunoichi Ninpōchō* (Tokyo, Kadokawa Shoten).

Yamada, Fūtarō (2014a). *Kōka Ninpōchō* (Tokyo, Kōdansha).

Yamada, Yūji. (2014c). *Ninja no Kyōkasho: Shin Mansenshūkai*. (Tokyo, Kasamashoin).

Yamada, Yūji (Ed.) (2014d). *The Ninja Book: the new Mansenshūkai* (Tsu City, Mie University) E-book.

Yamada, Yūji (2014e). *The Spirit of Ninja: A Study of the Global Ninja* Craze (Tsu City, Mie University) E-book.

Yamada, Yūji (2015a). *Ninja no Kyōkasho 2: Shin Mansenshūkai*. (Tokyo, Kasamashoin).

Yamada, Yūji (2015b). *Ninja Shugyō Manuaru*. (Tokyo, Jitsugyō no Nihon Sha).

Yamada, Yūji (2016). *Ninja no Rekishi*. (Tokyo, Kadokawa Shoten).

Yamada, Yūji (Ed.) (2017). *Ninja ninjutsu chō hiden zukan*. (Tokyo, Nagaoka Shoten).

Yamaguchi, M. (1963). *Ninja no Seikatsu*. (Tokyo, Yūzankaku).

Yamamoto, Taketoshi (2016). *Nihon no intelligence kōsaku* (Tokyo; Shinyosha).

Yokoyama, Takaharu. (1992). *Nobunaga to Ise.Iga - Mie Sengoku monogatari* (Tokyo, Sōgensha).

Yokoyama, Takaharu. (2006). *Iga Tenshō no ran* (Tokyo, Shinpu Shobō).

Yoshida Y. (Ed.) (1979). *Taikō ki* Vols 1, 3 and 4 (Tokyo, Kyōikusha).

Yoshimaru, Katsuya; Yamada, Yūji; Onishi, Yasumitsu *et. al.* (2014). *Ninja bungei kenkyū dokuhon* (Tokyo, Kasamashoin).

Yoshimaru, Katsuya (2014). 'From Shinobi to Ninja: Observations Focusing on the Tatsukawa Bunko Edition of Sarutobi Sasuke' in *The Truth about Ninja* (Briefing Notes for 2014) Mie University Faculty of Humanities, Law and Economics, p. 4.

Yoshimaru, Katsuya & Yamada, Yūji (2017). *Ninja no tanjō* (Tokyo, Kikkaku).

Yūki, S. (1988). 'Kōka jōkaku gun' *Rekishi Dokuhon Vol. 482 August 1988* pp. 114-121.

Index